Warm Ups Book

MIND Institute's Algebra Readiness Program

Warm Ups Book

MIND Institute's Algebra Readiness Program

MIND Institute

Warm Ups Book

MIND Institute's Algebra Readiness Program

Copyright © 2007 by MIND Institute

MIND Institute

www.mindinstitute.net

1503 South Coast Drive

Suite 202

Costa Mesa, California 92626

Printed in China

M1E1W03D07

ISBN: 1-933787-55-4

The MIND Institute is a nonprofit
research organization.

Contents

What are Warm Ups?

Warm Ups are included for each lesson in the MIND Institute's Algebra Readiness textbook: *A Blueprint for the Foundation of Algebra*. Warm Ups are mathematics exercises intended to be used at the beginning of a class period. They consist of 10 problems that are strategically designed to provide:

- practice and/or assessment of concepts and skills taught in the previous lesson;

- review of concepts and skills taught in earlier lessons; and

- preparation for the current lesson by reviewing necessary concepts and skills.

How are Warm Ups Used?

The Warm Ups can be used in a variety of ways, but should take no more than 10 minutes of the class period in order to have enough instructional time for the lesson. The following are some effective ways to implement the Warm Ups in the classroom.

- Allow five minutes for students to work on the Warm Ups on their own. Then use the remaining five minutes to check students' work as a class. Some options are:

  - Have individual students come to the front of the class and explain how they solved a particular problem. You may want to assign problems or call on students randomly.

  - Allow pairs of students to share their work.

  - Give students transparencies to write their solutions so they can be projected for the entire class to see.

  - The teacher may highlight certain problems and help students work through them.

  - Use the time to reteach selected problems based on prior assessment.

  - Allow teams of 3-4 students to work together to teach each other, checking their work and asking for help when needed.

- Use the Warm Ups as a quiz to assess student understanding.

- For a shorter Warm Ups exercise, assign only the five problems marked with a star (★). These problems focus on the most essential concepts and skills.

- On days when students are working on the computer, use some of the Warm Ups as small group instruction. It is often useful to review previously attempted problems and arrange students in groups according to instructional need. Due to time constraints, it is important to target the instruction efficiently and take no more than five minutes per group.

- Allow students to work cooperatively on the problems. They may be assigned groups, partners, or student selected groups. Be sure to monitor the class as they are working to assure that they are on task and can ask for help when needed.

The Warm Ups are extremely important as they provide ongoing assessment of the foundational and targeted standards of the Algebra Readiness program. They also provide repeated practice and recycling of concepts and skills needed for retention.

Name _____ Date _____ Period _____

Find the value of the variable.

★**1.** $9 + 7 = m$

2. $8 + 9 = p$

3. $4 + 8 + 3 = v$

Classify each as an equation or expression.

4. $d + f + 2$ _____

★**5.** 3 _____

★**6.** $8 + f = 3$ _____

7. n _____

Indicate whether the point is positive or negative.

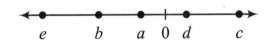

★**8.** a _____ b _____ c _____

d _____ e _____

Identify the whole numbers on the number line.

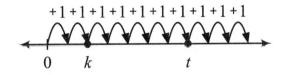

9. $k =$ _____

★**10.** $t =$ _____

Name _____ Date _____ Period _____

1. **Find the value of the expression** $3 + a$.

   a. when $a = 7$ **b.** when $a = 12$ **c.** when $a = 0$

★2. **Find the value of the expression** $d + e$.

   a. when $d = 9$ and $e = 6$ **b.** when $d = 0$ and $e = 4$ **c.** when $d = 7$ and $e = 6$

★3. **Find the missing values in the table of solutions.**

$$f + 4 = h + 1$$

f	h
0	
	7
5	
	10

Tell whether the values are a solution for the equation $r + 1 = u + 6$.

★4. $r = 5; \ u = 6$ ★5. $r = 8; \ u = 3$ 6. $r = 12; \ u = 7$

Find the solution to each equation.

★7. $6 + m = 14$ 8. $7 + n = 8$ 9. $3 + 3 + k = 12$

10. $f + 3 = 9 + 3$

Name _____ Date _____ Period _____

Simplify.

★**1.** $5 + v + 3 + 4$

2. $8 + c + 6 + d + 1 + 3$

3. Find the value of the expression $8 + f$.

    a. when $f = 9$

    b. when $f = 5$

    c. when $f = 3$

★**4.** Find the missing values in the table of solutions.

$$7 + e = g + 4$$

e	g
1	
	5
6	
	8

★**5.** Find the value of m in the equation $3 + m = 7 + n$.

    a. when $n = 4$

    b. when $n = 5$

    c. when $n = 37$

Find the solution to each equation.

    6. $m + 3 + 2 = 4 + 8$

    ★**7.** $d + d + d = 12$

    8. $8 = f + f + f + f$

    ★**9.** $5 + y = 7 + 5$

    10. $6 + m + 1 = 8 + 5$

Name _____ Date _____ Period _____

Use multiplication to express the repeated addition.

  1. $6 + 6 + 6 + 6 + 6 + 6 + 6$ ★**2.** $n + n + n$

  3.   **Find the value of the expression** $6 \times y$.

     a.   when $y = 7$

     b.   when $y = 9$

     c.   when $y = 4$

Use parentheses and multiplication to simplify each expression.

★**4.** $c + c + 3 + c + 3 + 3 + c + 3$ **5.** $1 + d + d + f + 1 + f + d + 1 + f$

Find the solution to each equation.

★**6.** $7 \times m = 0$ **7.** $11 = n + 6$

★**8.** $7 \times d = 7$ **9.** $12 + 12 + 12 = n \times 12$

★**10.** $w \times (3 + 5) = 3 + 3 + 3 + 5 + 3 + 5 + 5 + 5$

Name _____ Date _____ Period _____

Use multiplication to express the area of each rectangle. Then find the value of the expression.

★**1.**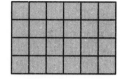

2.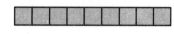

3.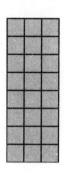

Solve.

★**4.** Mandy walked 4 blocks to the video store and 2 more blocks to the grocery store. She then walked 6 blocks home. Write an expression to describe how many blocks she walked in all. Find the value of the expression.

★**5.** Ramil caught three fish. Each fish was eight units long. How many units long would all three fish be if they were laid out head to tail?

Find the value of the variable.

6. $8 \times t = 32$

7. $8 + 4 + m = 7 + 8$

8. $3 \times w = 27$

★**9.** $48 = 6 \times y$

★**10.** $c + 1 = 4 + 8$

Name _____ Date _____ Period _____

Write an expression using multiplication and addition to describe the area of the shape composed of unit squares. Find the value of the expression in square units.

★1.

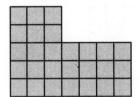

★2.

Find the value of the expression.

★3. $4 \times 8 + 3 \times 6$ 4. $5 \times (3 + 4)$ 5. $7 \times 6 + 3 \times 4$

Solve the equation.

★6. $8 \times (3 + 4) = t$ 7. $6 \times 9 = n + 40$

★8. $3 \times (s + 7) = 1 + 7 + 1 + 7 + 1 + 7$

Draw a shape with an area represented by each expression.

9. $3 \times (1 + 3)$ 10. $1 \times (3 + 4 + 5)$

Name _____ Date _____ Period _____

Define a variable to represent the unknown value and write an equation to describe each situation.

★**1.** Kelly scored 10 points. Mario scored 7 more points than Kelly did.

★**2.** When I earn three more dollars I'll have $22.

★**3.** We paid $12 for some hamburgers. Each hamburger cost $2.

4. Melissa earned $36 by working for several hours on Saturday. She gets paid $9 an hour.

5. There were eight slices of pizza. Jorge ate three slices last night and two this morning. I had one slice last night and ate the rest of the pizza this afternoon.

Rewrite using parentheses and multiplication.

6. $3 + e + e + 3 + 3 + e$

★**7.** $f + f + f + g + 2 + g + 2 + g + 2$

Write an expression for the area.

★**8.**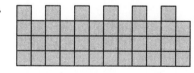

9.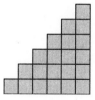

10. Define a variable for the unknown value and write an equation to describe the situation. Then solve the equation to answer the question.

There are 6 people at each table. If there are 42 people, how many tables are there?

Name _____ Date _____ Period _____

Use the distributive property to rewrite the expression as a single multiplication.

1. $3 \times f + 3 \times g$

★2. $5 \times 7 + 5 \times d + 5 \times e$

3. $k \times 2 + k \times m + k \times p$

★4. **Match the identities.**

a. $3 + m + n + d = 3 + n + m + d$

b. $3 \times m + 3 \times n + 3 \times d = 3 \times (m+n+d)$

c. $(3+4+m) \times 1 = (3+4+m)$

d. $3 + m + n + 0 = 3 + m + n$

i. additive property of 0

ii. multiplicative property of 1

iii. commutative property of addition

iv. distributive property

★5. **Use an identity to find the value of the variable. Which identity did you use?**

$$3 \times (5+3) = 3 \times a + 3 \times 3$$

Find the value of the variable in each equation.

★6. $a \times 8 = 0$

7. $3 \times 6 = c$

8. $d + 16 = 16$

★9. $f \times (8+6) = 8 + 6$

10. $4 + 7 + g = 8 \times 2$

Name _____ Date _____ Period _____

Write the value in expanded form.

1. 865

★2. 4,623

3. $10 + 10 + 10 + 1 + 1 + 1 + 1 + 1$

★4. $1000 + 1000 + 10 + 10 + 10 + 10 + 1$

Rewrite the expression in proper expanded form.

5. 24×10

★6. $7 \times 100 + 6 \times 100 + 3 \times 10 + 5 \times 1$

★7. $8 \times 100 + 4 \times 10 + 7 \times 10 + 8 \times 1 + 7 \times 1$

Use the distributive property to rewrite the expression as a single multiplication.

★8. $7 \times g + 7 \times h + 7 \times 3$

9. $m \times 3 + m \times p$

Solve for the variable.

10. $2 \times d + 6 \times 1 = 206$

Warm Up for Lesson 12

Name _____ Date _____ Period _____

Write the value of the expression in place value notation.

⋆**1.** $6 \times 1{,}000{,}000 + 7 \times 100{,}000 + 4 \times 1{,}000 + 6 \times 100 + 4 \times 10 + 9 \times 1$

2. $8 \times 10{,}000{,}000{,}000 + 6 \times 100{,}000{,}000 + 2 \times 1{,}000{,}000 + 5 \times 1{,}000 + 8 \times 10$

Write the number in expanded form.

⋆**3.** $6{,}350{,}700{,}003$

Write the number in standard form using place value.

⋆**4.** Seventy trillion, four billion, five hundred fifty-nine.

Write the number in words.

⋆**5.** $6{,}257{,}304{,}861$

Use the distributive property to rewrite the expression as a single multiplication.

6. $8 \times 4 + 8 \times 3$ **7.** $m \times 6 + m \times 3 + m \times 2$

8. $10 \times 5 + 10 \times 3$ **9.** $100 \times 3 + 100 \times 2 + 100 \times 1$

⋆**10.** $1{,}000 \times k + 1{,}000 \times p$

Name _____ Date _____ Period _____

Write the number in expanded form using exponents.

1. 612

★**2.** 400,303,900,675

Write the number in standard form using place value.

★**3.** $3 \times 10^9 + 7 \times 10^8 + 1 \times 10^5 + 6 \times 10^3 + 8 \times 10^2 + 4 \times 10^0$

Rewrite the number in proper expanded form.

★**4.** $5 \times 10^5 + 3 \times 10^7 + 3 \times 10^5$

5. $6 \times 10^8 + 7 \times 10^8 + 8 \times 10^3 + 8 \times 10^2 + 3 \times 10^2$

Write each number as 10 to a power.

6. 1   ★**7.** 1,000   8. 100,000

Find the sum.

9. $103 + 49$ ★**10.** $20,916 + 7,222$

Name _____ Date _____ Period _____

Rewrite in proper expanded form.

1. $36 \times 10 + 42 \times 1$ ⭐**2.** $48 \times 10 + 72 \times 1$

Rewrite the expression using the distributive property. Then write the expression in proper expanded form.

⭐**3.** $2 \times (7 \times 10 + 3 \times 1)$

4. $6 \times (4 \times 10 + 8 \times 1)$

Solve.

⭐**5.** A babysitting job pays $7 an hour. Last month you worked 26 hours. How much money did you earn?

⭐**6.** A package of pens costs $5. A package of pencils costs $4. How much will you pay for 2 packages of pens and one package of pencils?

Multiply.

7.
$$\begin{array}{r} 3\ 3 \\ \times\ \ \ 7 \\ \hline \end{array}$$

⭐**8.** 452×8

9. $6,305 \times 9$

Use repeated multiplication to show the meaning of the expression.

10. **a.** 10^2 **b.** 10^5

Name _____ Date _____ Period _____

A kilometer is 1,000 meters. Use this to find each measurement in meters.

⋆**1.** 75 kilometers

⋆**2.** 1,000 kilometers

Multiply to find the value of each expression.

⋆**3.** 150×10
150×100
$150 \times 1,000$
$150 \times 10,000$

⋆**4.** 8×7
80×7
80×700
$8,000 \times 7,000$

Draw a diagram, then write an equation to represent the situation. Solve the equation.

5. A football field is a rectangle, 360 feet long and 160 feet wide. The football team runs around the field to build endurance. If they run 10 laps around the field, how many feet are they running?

Write each number as 10 to a power.

6. One hundred million

7. One

Solve for the variable in each equation.

⋆**8.** $76 \times n = 76,000$

9. $d \times (60 + 3) = 600 + 30$

10. $576 + 9,081 = y$

Name _____ Date _____ Period _____

Write a rate for the following.

★**1.** From nickels to dollars

2. From \$3 to 2 pounds

For each situation, determine what units you are going FROM and what units you are going TO. Write the rate. Then perform the conversion to answer the question.

3. There are 24 hours in every day. How many hours are in 7 days?

★**4.** The school sells 5 dividers for \$2. How many dividers can you buy for \$10?

★**5.** There are 1000 meters in a km. How many meters are equivalent to 10 km?

Write in standard form using place value.

6. $7 \times 10^5 + 3 \times 10^3 + 2 \times 10^2 + 7 \times 10^1 + 8 \times 10^0 + 9 \times 10^0$

Write in words.

★**7.** 625,000,503

Find the value of each expression.

8. 235×9

9. $\begin{array}{r} 4\ 4\ 4 \\ \times\quad\ 6 \\ \hline \end{array}$

★**10.** $5 \times 64 + 7 \times 380$

Name _____ Date _____ Period _____

Find the conversion rate.

★1. From minutes to hours

2. From dimes to pennies

Find the rate for each situation. Then answer the question.

★3. Only one out of four students walks to school. If there are 28 students in the class, how many students walk to school?

★4. Four oranges cost $3. How much will a dozen oranges cost?

5. One basketball player makes 6 out of 10 free throw shots. If he shoots 30 shots, how many will he make?

Use repeated multiplication to show the meaning of the expression.

6. 10^5

Use repeated addition to show the meaning of the expression.

★7. $3 \times (m + 4)$

Write in standard form using place value.

8. $4 \times 10^5 + 3 \times 10^2 + 5 \times 10^1 + 8 \times 10^0$

Find the value of the expression.

★9. $73,245 + 7,941$

10. $8,203 \times 6$

Name _____ Date _____ Period _____

Find the ratio for the input-output machine.

1. $6 \times 10 \longrightarrow$ [?] $\longrightarrow 5 \times 10$

Find the output for each machine.

2. $16 \longrightarrow$ [$\frac{7}{8}$] $\longrightarrow$?   ★3. $45 \longrightarrow$ [$\frac{2}{3}$] $\longrightarrow$?   ★4. $24 \longrightarrow$ [$\frac{1}{12}$] $\longrightarrow$?

Solve.

5. Divide 42 into 6 parts by solving for k: $42 = 6 \times k$

★6. Divide 54 into 9 parts by solving for t: $54 = 9 \times t$

Use the rate to perform the conversion.

7. Convert 15 mids to fods using this rate: $\frac{3 \text{ fods}}{5 \text{ mids}}$

★8. Convert 24 jigs to jags using this rate: $\frac{8 \text{ jags}}{6 \text{ jigs}}$

Find the rate for each situation. Then answer the question.

★9. Marcos answered 3 multiplication facts in 10 seconds. At that rate how many seconds will it take to do 15 multiplication facts?

10. Hair clips come four to a package. Each package sells for $7. How many hair clips can be purchased for $42?

Name _____ Date _____ Period _____

Find the value of each expression.

★**1.** $\frac{3}{8} \times 48 =$

2. $\frac{6}{5} \times 45 =$

★**3.** $\frac{7}{1} \times 8 =$

Use the distributive property to rewrite the expression. Then find the value of the expression.

★**4.** $\frac{7}{4} \times 12 + \frac{7}{4} \times 8$ **5.** $\frac{1}{5} \times 7 + \frac{1}{5} \times 8$

Solve for the variable k in each equation.

★**6.** $\frac{k}{3} \times 24 = 16$ **7.** $\frac{3}{5} \times k = 9$

Find the value of each expression.

8. $\begin{array}{r} 1\,7\,0\,0\,9 \\ \times \quad\quad 7 \\ \hline \end{array}$ **9.** $867 \times 1{,}000 + 3 \times 1{,}000$

★**10.** $4{,}716 + 985$

Name _____ Date _____ Period _____

Find the value of the expressions. Don't forget the units.

1. $\dfrac{6 \text{ feet}}{10 \text{ seconds}} \times 60 \text{ seconds}$

★2. $\dfrac{\$12}{5 \text{ pounds}} \times 20 \text{ pounds}$

Fill in the blanks with the missing units.

★3. $\dfrac{5\,\boxed{}}{2\,\boxed{}} \times 14 \text{ pages} = 35 \text{ faces}$

Define a variable to represent the unknown value and write an equation to describe the situation. Then solve the equation to answer the question.

★4. Three pounds of bananas sell for $2. How many pounds can be purchased for $6?

★5. The SUV gets 9 miles per gallon. How many gallons of gas are needed to go 90 miles?

Find the value of the variable.

6. $8 \times 28 = a$

★7. $7 \times c = 6 \times 10^1 + 3 \times 10^0$

8. $685 \times 4 = f$

9. $n \times 8 = 52 + 4$

10. $41 + 13 = m \times 6$

Name _____ Date _____ Period _____

Rewrite each division expression as multiplication by the inverted rate or ratio. Find the value.

★1. $63 \div \frac{7}{5}$

★2. 48 minutes $\div \dfrac{6 \text{ minutes}}{7 \text{ pages}}$

Rewrite each multiplication expression as division. Find the value.

3. $\frac{1}{7} \times 63$

★4. $\dfrac{1 \text{ hour}}{\$8} \times \$24$

Find the value of the variable.

★5. $a = 20 \div 4$ **6.** $5 \times m = 50$

7. $t \div 3 = 5$ **8.** $\frac{3}{4} \times d = 12$

9. $15 \div \frac{2}{3} = f \times 15$ **★10.** $\frac{7}{8} \times w = 0$

Name _____ Date _____ Period _____

Write an expression to answer the question. Find the value of the expression.

★**1.** What is the quotient when 63 is divided by 9?

2. There are 5 boxes of 144 pencils. How many pencils are there in all 5 boxes?

What is the area of the rectangle? Name 4 different factors of this area.

★**3.**

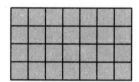

Find all the factors of each number.

★**4.** 30

5. 19

Rewrite the division expression as multiplication by the inverted ratio. Find the value of the expression.

★**6.** $72 \div \frac{9}{7}$

7. $48 \div \frac{6}{7}$

Answer each question.

★**8.** Is 10 divisible by 5?

9. Is 24 a factor of 6?

10. Is $50 \div 25$ a whole number?

Name _____ Date _____ Period _____

Solve for the remainder r in each equation.

1. $8 \times 8 + r = 66$

★**2.** $3 \times 31 + r = 94$

★**3.** Find the value of each expression.

$24 \div 6$

$2{,}400 \div 6$

$240{,}000 \div 6$

Find the value of the variable in each equation.

★**4.** $4 \times 8 + g = 34$

5. $6 \times 4 + a = 27$

$6 \times 40 + b = 270$

$6 \times 400 + c = 2{,}700$

Find the value of each expression.

★**6.** $3{,}275 \times 7$

7. $286 + 7$

Rewrite each division as multiplication by the inverted rate. Then find the value of the expression.

8. $63 \text{ marbles} \div \dfrac{9 \text{ marbles}}{4 \text{ players}}$

★**9.** $36 \text{ monkeys} \div \dfrac{9 \text{ monkeys}}{5 \text{ acres}}$

10. $24 \text{ feet} \div \dfrac{2 \text{ feet}}{3 \text{ years}}$

Name _____ Date _____ Period _____

Find the difference.

★**1.**   703
       − 238
       _____

2.   4327
      − 865
      _____

Find the value of the variable.

★**3.** $26 + c = 938$

4. $409 + d = 826$

Define a variable, write an equation, and solve.

★**5.** The coastal redwoods can grow to a height of 367 feet tall and the saguaro cactus can be 53 feet tall. What is the difference between these two heights?

Divide. Then write an equation that expresses the dividend as a product of whole numbers plus a remainder.

6. $84 \div 4$

7. $72 \div 3$

8. $61 \div 5$

★**9.** $52 \div 6$

★**10.** $85 \div 7$

Name _____ Date _____ Period _____

Rewrite using the division symbol, ÷ .

1. $5 \overline{)4601}$

Rewrite using the long division symbol, $\overline{)}$.

★2. $84 \div 6$ **★3.** $5 \div 9$

Divide.

★4. $882 \div 7$ **5.** $11{,}112 \div 8$

Rewrite the division expression as multiplication by the inverted ratio. Then find the value of the expression.

★6. $6 \div \frac{6}{12}$ **★7.** $48 \div \frac{6}{8}$ **8.** $24 \div \frac{4}{3}$

Solve.

9. There were 76 tickets sold on Thursday and 82 tickets sold on Friday. How many more tickets were sold on Friday than Thursday?

10. The tickets cost $15 each. How much will it cost to buy 7 tickets?

Name _____ Date _____ Period _____

Find the value of each expression. Include the units in your answer.

★**1.** $\dfrac{3 \text{ books}}{5 \text{ days}} \times 365 \text{ days}$

2. $\dfrac{3 \text{ quarts}}{2 \text{ pounds}} \times 92 \text{ pounds}$

Define a variable to represent the unknown value. Write an equation to describe the situation. Then solve the equation to answer the question.

3. I am planning a trip to New York City. It is 3,943 kilometers from Los Angeles to New York City. From San Francisco to New York City it is 4,139 kilometers. How many fewer kilometers will I travel if I leave from Los Angeles?

★**4.** It takes 23 minutes to drive 7 miles to school in the morning. After class in the evening it only takes 12 minutes to get home. What is the average speed for the round trip?

★**5.** The cafeteria orders 3 cartons of regular milk for every 2 cartons of chocolate milk. If they ordered 174 regular milk cartons, how many cartons of chocolate milk will they receive?

Rewrite the repeated addition as multiplication.

★**6.** $1 + 1 + 1 + 1 + 1$

7. $6 + 6 + 6 + 6 + 6 + 6$

Solve for the variable.

8. $685 + 399 = k$

9. $m = 471 \times 2$

★**10.** $y = 668 \div 4$

Name _____ Date _____ Period _____

Write each expression as a fraction.

★**1.** $\frac{1}{5} + \frac{1}{5} + \frac{1}{5} + \frac{1}{5} + \frac{1}{5} + \frac{1}{5} + \frac{1}{5}$

★**2.** $6 \times \frac{1}{3}$

Rewrite the fraction as the product of a whole number and fraction with numerator 1.

★**3.** $\frac{3}{5}$

4. $\frac{18}{4}$

Use the symbol >, < or = to make the statement true.

★**5.** $4 \,\square\, \frac{4}{4}$

Find the value of the expression.

6. $\frac{2}{3} \times 6$

7. $\frac{7 \text{ jobs}}{4 \text{ days}} \times 24 \text{ days}$

8. Find the length of the missing side if the area of the rectangle is 27 square units.

Find the area.

★**9.**

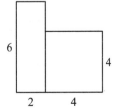

10.

Name _____ Date _____ Period _____

Here $\frac{9}{a}$ of the circles are black. What is a?

1.

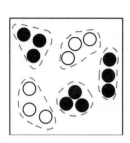

Here $\frac{12}{b}$ of one circle is shaded. What is b?

★**2.**

Find the value.

★**3.** $\frac{3}{5}$ of $100

Answer each question. Use an equation to represent the result.

4. $\frac{5}{9}$ of the students prefer pepperoni pizza to cheese pizza. If there were 621 students surveyed, how many students choose pepperoni pizza?

★**5.** The clothing store is having a $\frac{1}{3}$ off sale. Monique chooses a dress that was originally priced $84.
   a. How much will she save?
   b. What is the sale price of the dress?

Rewrite using the distributive property. Then find the value.

★**6.** $\frac{3}{8} \times 20 + \frac{3}{8} \times 4$

7. $\frac{1}{3} \times 8 + \frac{1}{3} \times 7$

Rewrite in proper expanded form.

8. $6 \times 10^3 + 2 \times 10^3 + 4 \times 10^1 + 3 \times 10^1 + 8 \times 10^0$

Solve for the variable.

★**9.** $j = 6{,}042 \div 6$

10. $z = 406 \times 8$

Name _____ Date _____ Period _____

Rewrite the expression using the commutative property of multiplication.

★**1.** 38×8

Rewrite the expression as a fraction.

★**2.** $5\overline{)4}$

3. $16 \div 3$

Rewrite the division as multiplication by the inverted rate.

★**4.** $6 \div 4$

5. $8\overline{)36}$

Find the value of the variable.

6. $6 \times 10 = 10 \times n$

★**7.** $\frac{3}{2} \times (3 + y) = \frac{3}{2} \times 7$

★**8.** $m = \frac{2}{3} \times 15$

9. $6 \div \frac{1}{5} = 6 \times d$

10. $7 \times m = 448$

Name _____ Date _____ Period _____

What fraction of each shape is shaded gray?

⋆**1.**
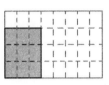

⋆**2.**
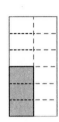

Multiply.

⋆**3.** $\frac{7}{5} \times \frac{3}{7}$

4. $\frac{8}{3} \times 7$

5. $\frac{2}{9} \times \frac{1}{8}$

Solve for the variable.

⋆**6.** $\frac{5}{a} \times \frac{7}{1} = \frac{35}{8}$

7. $\frac{3}{8} \times \frac{m}{6} = \frac{27}{48}$

Write the number in standard form using place value.

8. $5 \times 10^6 + 4 \times 10^5 + 2 \times 10^2 + 8 \times 10^0$

Rewrite in proper expanded form.

9. $6 \times 10^4 + 5 \times 10^3 + 7 \times 10^3 + 2 \times 10^2 + 7 \times 10^1 + 1 \times 10^0$

⋆**10.** Mimi kept track of how long it took her to do her homework last night. The first 10 problems took her 7 minutes to complete. The second 10 took 12 minutes and the last 10 took 21 minutes. What is her average rate in problems per minute?

Name _____ Date _____ Period _____

Find a fraction that is equivalent to the given fraction.

★1. $\dfrac{2}{3}$

2. $\dfrac{1}{5}$

Using the denominators given, create 3 equivalent fractions.

3. $\dfrac{3}{8}, \dfrac{\square}{24}, \dfrac{\square}{48}, \dfrac{\square}{64}$

★4. $\dfrac{7}{6}, \dfrac{\square}{24}, \dfrac{\square}{30}, \dfrac{\square}{54}$

Define a variable, write an equation and solve.

★5. A pizza was cut in fourths. Then each fourth was cut in 3 pieces. If Mark ate two pieces, what fraction of the pizza did he eat?

★6. A quilt is made up of 80 squares. Each square has an area of 8 square inches. What is the total area of the quilt?

Find the value of each expression.

7. $\dfrac{3}{6} \times \dfrac{8}{7}$

8. $48{,}703 \times 6$

9. $5 \div \dfrac{1}{6}$

★10. $\dfrac{282}{6}$

Name _____ Date _____ Period _____

Rewrite the division as multiplying by the inverted fraction.

★1. $\dfrac{7}{5} \div \dfrac{3}{4}$

Divide.

★2. $\dfrac{8}{7} \div \dfrac{5}{3}$

3. $\dfrac{21}{5} \div \dfrac{10}{3}$

Use division to compare fractions. Place the symbol $>$, $<$ or $=$ to make the statement true.

★4. $\dfrac{5}{12} \bigcirc \dfrac{3}{8}$

5. $\dfrac{4}{6} \bigcirc \dfrac{6}{9}$

Find the value if $n = 5$ and $m = 7$.

6. $\dfrac{3}{5} \div \dfrac{n}{8}$

★7. $\dfrac{m}{n} \div \dfrac{8}{6}$

8. $\dfrac{13}{m} \times \dfrac{n}{24}$

Find the value of the expression. Remember to include the units in your answer.

★9. $\dfrac{5 \text{ apples}}{3 \text{ students}} \times 255 \text{ students}$

10. $\dfrac{17 \text{ hours}}{5 \text{ pages}} \times 45 \text{ pages}$

Name _____ Date _____ Period _____

★1. Find the value of the expression.

$$\frac{7 \text{ dollars}}{2 \text{ pounds}} \times 94 \text{ pounds}$$

★2. Use the result from question 1 to generate an equivalent rate.

$$\frac{7 \text{ dollars}}{2 \text{ pounds}} = \frac{\boxed{}}{94 \text{ pounds}}$$

★3. Test to see if the rates are equal. Use = or ≠ to make the statement true.

$$\frac{6 \text{ days}}{4 \text{ hours}} \bigcirc \frac{15 \text{ days}}{10 \text{ hours}}$$

★4. Kim earns two dollars every 3 minutes. How much will Kim earn in 60 minutes? Write the result as an equivalent hourly rate.

5. If Kim works 40 hours per week, write the result as an equivalent weekly rate.

6. Is 3 a factor of 24?

7. Is 56 ÷ 9 a whole number?

★8. Is 27 ÷ 5 a whole number?

9. What is the remainder when 46 is divided by 3?

10. Solve for the variable:

$$5 \times n = 35$$

Name _____ Date _____ Period _____

★**1.** Categorize each number as even or odd: 24, 38, 57, 89, 254, 150.

★**2.** Use division to see if 66 is divisible by 4.

3. List 5 multiples of 7.

★**4.** There are 36 students in the class. Can the students be divided equally in groups of 4? Groups of 5? Groups of 6? Group of 8?

5. There are 86 students going on a field trip. They are required to be in groups of at least 4 but no more than 5 students. Is it possible for all of the groups to be the same size? Explain your answer.

Find the value of each expression.

6. $3020 \div 5$

★**7.** $8 \times \frac{2}{3} + 4 \times \frac{2}{3}$

8. $\begin{array}{r} 5\,0\,2\,6 \\ -\ 2\,1\,3\,4 \\ \hline \end{array}$

★**9.** $\frac{5}{7} \div \frac{6}{7}$

10. $\frac{2}{3} \times \frac{8}{5}$

Name _____ Date _____ Period _____

★1. Identify the numbers as prime or composite.

    2 –

  14 –

    9 –

2. Identify the numbers as prime or composite.

  17 –

  23 –

  49 –

★3. Write the first 10 prime numbers.

4. List the first five multiples of 9.

5. Use division to see if 2,186 is divisible by 3.

★6. Find all the factors of 36.

Find the value of each expression.

★7. $\dfrac{8260}{5}$

★8. $\dfrac{5}{3} \div \dfrac{8}{17}$

9. 4280×9

10. $\dfrac{5}{18} \times \dfrac{7}{19}$

Name _____ Date _____ Period _____

★1. State whether or not each number is divisible by 3.

    a. 518   **b.** 2,955   **c.** 87

★2. State whether or not each number is divisible by 5.

    a. 200   **b.** 523   **c.** 100,530

★3. State whether or not each number is divisible by 9.

    a. 213   **b.** 555   **c.** 306

4. State whether or not each number is divisible by 9.

    a. 180   **b.** 783   **c.** 512

5. **Put a digit in the ones place to make the number divisible by 9.**

    512,60__

Write the number in expanded form using exponents.

6. 512   ★7. 7,263

8. Write the first 10 prime numbers.

★9. Identify the number as prime or composite.

    7 –

    9 –

  27 –

  39 –

10. Is 236 divisible by 6? How can you tell without dividing?

Name _____ Date _____ Period _____

★1. Label the numbers as prime, composite, or neither.

1 –   21 –   43 –

2 –   23 –   59 –

9 –   39 –   87 –

2. Write the first 10 prime numbers.

★3. Rewrite the prime factorization so the factors are ordered from smallest to largest. What number has this prime factorization?

$5 \times 2 \times 3 \times 2$

4. Write the prime factorization for 48.

5. Write the prime factorization for 234.

★6. Write the prime factorization for 100.

★7. Write the prime factorization for 1,000.

Solve each equation.

8. $36 \times n = 360$ **9.** $700 \div 7 = n$ **★10.** $4,800 = n \times 48$

Name _____   Date _____   Period _____

Find the value.

★**1.** $5,200 \times \frac{1}{10}$

★**2.** $8,200,000 \div 100$

3. $\frac{1}{1000} \times 712,000$

Write the prime factorization for each number.

4. 800

★**5.** 450

6. 128

Write each number in expanded form using exponents.

7. 10,029

★**8.** Forty-three million, eighty-seven thousand, twenty-nine.

Solve each problem.

★**9.** We bought 2 pies for $15. How much will it cost to buy 18 pies?

10. For the art project, 8 boxes of watercolor pencils are needed. Each box contains 24 pencils. How many watercolor pencils are there in all?

Name _____ Date _____ Period _____

Write the prime factorization of each number using exponents.

★1. 88   **★2.** 120

Use prime factorization to find the value.

3. $\sqrt{225}$

Determine if the following are perfect squares.

★4. $8{,}100 = 2 \times 2 \times 3 \times 3 \times 3 \times 3 \times 5 \times 5$ **5.** $9{,}000 = 2^3 \times 3^2 \times 5^3$

6. Express the exponents as repeated multiplication.

$3^2 \times 5^1 \times 7^3$

Define a variable and represent the situation with an equation. Then solve the equation to answer the question.

7. The school is buying a set of 5 small erasers for 8¢. How much will it cost to buy 720 erasers?

★8. Lined filler paper costs $3 for 500 sheets of paper. How many sheets of paper can I buy for $150?

Rewrite each expression using the distributive property. Then find the value of the expression.

9. $\frac{3}{2} \times 5 + \frac{3}{2} \times 3$ **★10.** $\frac{5}{3} \times 5 + \frac{5}{3} \times 10$

Name _____ Date _____ Period _____

Write the numerator and denominator in prime factored form. Use the factorization to simplify the fraction.

★**1.** $\dfrac{24}{18}$

2. $\dfrac{30}{32}$

★**3.** Express the exponents as repeated multiplication, then simplify the expression.

$$\dfrac{2^2 \times 3^1}{2^1 \times 3^2 \times 5}$$

Use fraction multiplication to solve the rate problems.

★**4.** At the bake sale, the soccer team was selling 2 cakes for $15. At that rate, how much should they charge for $\frac{1}{3}$ of a cake?

5. At the same sale, the team sold 8 ounce glasses of lemonade for 20¢. When they ran out of glasses, they had to use 6 ounce glasses. Using the same rate, how much should they charge for the smaller size?

Find the value of each expression. Simplify if possible.

★**6.** $3 \times \dfrac{6}{15}$ **7.** $\dfrac{8}{9} \times \dfrac{5}{12}$ **8.** $6 \div \dfrac{9}{7}$

★**9.** $\dfrac{5}{3} \div \dfrac{15}{2}$ **10.** $\dfrac{8}{15} \times \dfrac{12}{5}$

Name _____ Date _____ Period _____

Put the denominator of each fraction in prime factored form. Then rewrite the fractions with a common denominator.

★**1.** $\frac{7}{10}$ and $\frac{5}{8}$

★**2.** $\frac{4}{9}$ and $\frac{1}{12}$

Rewrite the fractions with a common denominator. Compare the fractions using >, <, or = .

★**3.** $\frac{8}{15}$ ◯ $\frac{7}{12}$

4. $\frac{10}{6}$ ◯ $\frac{12}{9}$

★**5.** For the upcoming Super Bowl, the grocery store was selling 12 cans of soda for $5. Another store was selling 18 cans for $7. Which store had the better price?

Divide. Write your answer with remainders.

6. $527 \div 4$

7. $3,207 \div 8$

★**8.** $9 \overline{)128}$

9. $3 \overline{)815}$

10. $7,250 \div 5$

Name _____ Date _____ Period _____

Add the fractions. Use prime factorization to find the common denominator if it is needed.

★1. $\frac{7}{9} + \frac{2}{3}$ ★2. $\frac{9}{5} + \frac{4}{5}$

3. $\frac{3}{4} + \frac{7}{6}$

Add. Express the answer as a mixed number.

★4. $\frac{5}{17} + 6$ 5. $6 + \frac{6}{7} + 3$

Define a variable, write an equation, and solve.

★6. Markum Middle School's colors are red and blue. On spirit day, $\frac{1}{4}$ of the students wore blue and $\frac{2}{3}$ of the students wore red. The rest were not in school colors. What fraction of the students wore school colors?

Find the value of each expression. Simplify if possible.

7. $\frac{2}{3} \times \frac{7}{8}$ 8. $\frac{5}{7} \div \frac{6}{5}$

★9. $6 \times \frac{5}{12}$ 10. $15 \div \frac{4}{3}$

Name _____ Date _____ Period _____

Express the fractions as mixed numbers.

★**1.** $\dfrac{15}{8}$

2. $\dfrac{53}{9}$

Express the mixed numbers as a single fraction.

3. $2\dfrac{1}{2}$

★**4.** $8\dfrac{6}{8}$

Write the fraction or mixed number.

★**5.** Thirty-two fifths

★**6.** One and one half

Find the value of each expression. Simplify if possible.

7. $\dfrac{5}{7} \times \dfrac{2}{8}$

★**8.** $\dfrac{5}{3} + \dfrac{7}{6}$

9. $\dfrac{5}{7} \div \dfrac{6}{4}$

10. $\dfrac{7}{8} + \dfrac{1}{6}$

Name _____ Date _____ Period _____

Solve for the variable.

1. $\frac{3}{8} + f = \frac{9}{8}$

2. $k + \frac{1}{2} = \frac{3}{4}$

3. $\frac{5}{8} + m = \frac{13}{4}$

★4. $2\frac{1}{3} + v = 5\frac{1}{6}$

★5. Carla has read $\frac{5}{6}$ of the book. The teacher said she has until Friday to read $\frac{9}{10}$ of the book. How much of the book must she read before Friday?

6. The school store is having a sale on booklets of lined paper. They are selling 8 booklets for $6. How many booklets can I buy for $9?

★7. Felt costs $3 per yard. We need $\frac{3}{4}$ of a yard of felt for the project. How much will it cost?

8. Write 521,003 in expanded form using exponents.

Find the value of each expression.

★9. $\frac{3}{5} + \frac{4}{20}$

★10. $\frac{1}{2} + \frac{1}{5}$

Name _____ Date _____ Period _____

For each value, find an equivalent fraction that has a denominator of 10.

1. $\frac{17}{2}$

⋆**2.** $\frac{8}{5}$

⋆**3.** 8

⋆**4.** $\frac{70}{100}$

5. $\frac{30}{20}$

Solve for the variable.

6. $\frac{5}{8} \times 5 = c$

⋆**7.** $2\frac{4}{5} + 5\frac{4}{15} = m$

⋆**8.** $8 \div \frac{5}{7} = w$

9. $3\frac{3}{4} + y = 6\frac{1}{4}$

10. $\frac{5}{3} + \frac{7}{2} = a$

Name _____ Date _____ Period _____

Find an equivalent fraction with a denominator of 100.

1. $\frac{7}{2}$

★2. $\frac{14}{25}$

Express the fraction as a percent.

★3. $\frac{17}{20}$

★4. 2

Rewrite the number using fraction notation. Then simplify the fraction.

★5. 76%

6. 150%

★7. Convert 225 centimeters to meters.

8. Convert 12 decimeters to centimeters.

Write the numbers in prime factored form using exponents.

9. 1000

10. 5,000

Name _____ Date _____ Period _____

Write the fraction with a denominator of 1000.

1. $\dfrac{5}{100}$

★2. $\dfrac{42}{250}$

Find the value of the variable.

★3. $\dfrac{51}{1000} + \dfrac{825}{1000} = h$

4. $\dfrac{4}{10} + \dfrac{5}{100} = n$

Approximate each fraction to the nearest percent.

★5. $\dfrac{5}{7}$

6. $\dfrac{2}{3}$

Express the value using fraction notation. Then simplify the fraction.

7. 32%

★8. 200%

Express the fraction as a percent.

★9. $\dfrac{3}{20}$

10. $\dfrac{1}{25}$

Name _____ Date _____ Period _____

Use prime factorization to simplify each fraction. Determine if it can be expressed exactly as a decimal fraction.

⋆**1.** $\dfrac{6}{16}$

2. $\dfrac{21}{12}$

Write the fraction in expanded form.

3. $\dfrac{76}{100}$

⋆**4.** $\dfrac{246}{1000}$

Rewrite the percentage in expanded form using fractions.

5. 81%

⋆**6.** 236%

Find the value of the expression. Write the value as a simplified fraction.

⋆**7.** $\dfrac{4}{10} + \dfrac{8}{100} + \dfrac{6}{1000}$

8. $\dfrac{1}{100} + \dfrac{6}{1000} + \dfrac{4}{10}$

Find the equivalent decimal fraction for each number.

9. $\dfrac{33}{200}$

⋆**10.** $\dfrac{9}{4}$

Name _____ Date _____ Period _____

Use decimal place value to write the value of each expression.

1. $\dfrac{5}{10} + \dfrac{9}{1000}$

★**2.** $6 \times 100 + 3 \times 1 + \dfrac{7}{100}$

Write the number in decimal notation.

★**3.** Thirty-nine thousandths.

4. Seven hundredths.

Express the percent as a fraction and as a decimal.

★**5.** 5.36%

6. 168%

Express the fraction as a decimal and as a percent.

★**7.** $\dfrac{3}{4}$

8. $\dfrac{19}{10}$

Write the number in expanded form.

★**9.** 40.104

10. $\dfrac{520}{1000}$

Name _____ Date _____ Period _____

Write the decimal in expanded form. Then write the value of the expression as a decimal fraction.

⋆**1.** 3.02

2. 0.98

Write the decimal in fraction notation. Simplify the fraction.

⋆**3.** 0.003

4. 1.6

Write the number in decimal notation.

⋆**5.** Seven hundred eight and sixty-two thousandths.

Write the decimal in words.

6. 3,201.04

Express the decimal as a fraction and percent.

⋆**7.** 0.6

Express the fraction as a decimal and as a percent.

⋆**8.** $\frac{8}{2}$

Rewrite using the distributive property.

9. $6 \times (m + 4)$

10. $3 \times \frac{2}{5} + 4 \times \frac{2}{5}$

Name _____ Date _____ Period _____

Add.

★1. $2.7 + 14.56$

2. $.64 + 17.009$

Find the value of the variable.

★3. $4.08 + p = 17.2$

4. $b + .39 = 1$

★5.

6.

Solve.

★7. The counter was 2.73 meters long. The stove uses 0.76 meters of the counter. How many meters long is the rest of the counter?

Find the value of the expression.

8. $6{,}258 \times 7$

★9. $\dfrac{3}{10} + \dfrac{5}{100} + \dfrac{2}{10}$

10. 6×10^3

Name _____ Date _____ Period _____

Multiply.

★**1.** 0.4×0.5

2. 0.027×0.8

Write the percent as a decimal fraction.

3. 5%

★**4.** 326%

Solve.

★**5.** The book store offers a 10% student discount. A student bought a notebook for $3.96, a pencil for 65¢, and dividers for $2.59. With the discount, how much will the student save (before tax)?

6. How much is 8% of $76.50?

Rewrite using the distributive property.

7. $\frac{2}{3} \times 7 + \frac{2}{3} \times 5$

★**8.** $7 \times \left(\frac{1}{10} + \frac{1}{100} \right)$

Write each number as a percent.

9. $\frac{1}{20}$

★**10.** $\frac{1}{25}$

Name _____ Date _____ Period _____

Multiply.

★1. 622×2.5

2. 467.8×0.1

3. 1.06×0.31

Find the area of the shape.

★4.

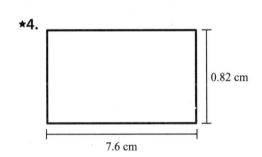

0.82 cm

7.6 cm

5.

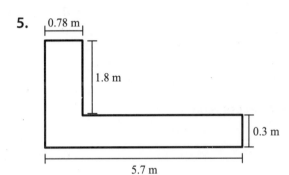

0.78 m

1.8 m

0.3 m

5.7 m

Write the percent as a fraction.

6. 3%

★7. 127%

Find the value of each expression.

★8. 5% of $\$185.20$

9. 9% of $\$42.10$

★10. $\dfrac{3 \text{ centimeters}}{11 \text{ hours}} \times 88 \text{ hours}$

Name _____ Date _____ Period _____

For each rate, find the equivalent unit rate. Express the rate in decimal notation.

⋆**1.** $\dfrac{7 \text{ dollars}}{2 \text{ hamburgers}}$

2. $\dfrac{18 \text{ words}}{5 \text{ inches}}$

Convert from inches to centimeters. There are exactly 2.54 centimeters in an inch.

⋆**3.** 17.1 inches

4. 6 inches

Find the value of each expression.

5. $\dfrac{4}{5} \times \dfrac{2}{3}$

6. $1\dfrac{2}{15} + 3\dfrac{1}{9}$

⋆**7.** $\dfrac{28}{5} \div \dfrac{7}{10}$

⋆**8.** $8{,}064 \div 3$

⋆**9.** $60{,}816 \div 8$

10. $\dfrac{12.2 \text{ ft}}{\text{second}} \times 90 \text{ seconds}$

Name _____ Date _____ Period _____

Divide. Check your work by multiplying.

1. $26.08 \div 8$

★2. $48.72 \div 5$

Solve for the variable.

★3. $k = 78.3 \div 4$

4. $m = 653 \div 5$

Find the unit rate. Include units in your answer.

★5. $\dfrac{6.38 \text{ meters}}{4 \text{ minutes}}$

Rewrite each fraction as a decimal and as a percent.

★6. $\dfrac{17}{20}$ **7.** $\dfrac{38}{50}$

★8. $\dfrac{9}{5}$ **9.** $\dfrac{27}{25}$

10. $\dfrac{3}{4}$

Name _____ Date _____ Period _____

Change the fractions to a decimal using long division. Use bar notation if it is a repeating decimal.

★**1.** $\frac{5}{6}$ 　　　　　　★**2.** $\frac{28}{5}$ 　　　　　　**3.** $\frac{7}{8}$

Approximate each decimal to 6 decimal places.

★**4.** $\frac{15}{9}$

5. $4.0\overline{314}$

6. Classify each of the numbers in questions #1–5 as a terminating or repeating decimal.

Without doing any calculations, determine if the fraction will be a terminating or repeating decimal. Explain your answer.

7. $\frac{7}{8}$ 　　　　　　　　★**8.** $\frac{100}{3}$

Find the value of each expression.

★**9.** $56 \div \frac{1000}{2}$ 　　　　　　**10.** $13.26 \div 6$

Name _____ Date _____ Period _____

Find the value of each expression.

★**1.** 3.26×10
3.26×100
3.26×1000

2. 0.409×10
0.409×100
0.409×1000

Generate equivalent expressions.

★**3.** $\dfrac{17.103}{0.008} = \dfrac{?}{0.8} = \dfrac{17,103}{?}$

Generate equivalent expressions so that the divisor is a whole number and then divide.

4. $42.3 \div 0.08$

5. $714 \div 0.006$

★**6.** $0.05 \overline{\smash{)}0.3}$

Write the fraction as a decimal and as a percent.

7. $\dfrac{5}{4}$

8. $\dfrac{130}{200}$

Solve for the variable.

★**9.** $7 \div \dfrac{2}{3} = v \times \dfrac{3}{2}$

★**10.** $\dfrac{2}{3} \times 5 + \dfrac{2}{3} \times 7 = \dfrac{2}{3} \times m$

Name _____ Date _____ Period _____

Write an equivalent expression in which the divisor is a whole number and then divide.

★**1.** $14.7 \div 0.05$

★**2.** $1.728 \div 0.003$

Simplify the fraction.

★**3.** $\dfrac{18}{24}$

★**4.** $\dfrac{560}{5600}$

Use the additive property of 0 to write an equivalent expression. Find the value.

★**5.** $13 + 6.25$

Match the property used by each statement to the name of the property.

6. $\dfrac{3}{7} \times 18 + \dfrac{3}{7} \times 3 = \dfrac{3}{7} \times (18 + 3)$

    a. commutative property of multiplication

7. $6 + a + 42.1 = a + 6 + 42.1$

    b. associative property of multiplication

8. $7.6 \times (5.2 \times 10) = (7.6 \times 5.2) \times 10$

    c. distributive property

9. $0 \times m \neq 6$

    d. commutative property of addition

10. $(a \times b) \neq (b \times a) + 1$

    e. multiplicative property of 0

Name _____ Dàte _____ Period _____

Apply the commutative property to the part of the expression indicated by the $\downarrow$.

1.
$$\downarrow$$
$$\frac{7}{4} \times 6 + 7 \times 32$$

★2.
$$\downarrow$$
$$(m+8) \times (4+5)$$

Put parentheses around each term of the expression.

★3. $5 \times 0.6 \times p + 3 \times f + f$

4. $a \times b + 6 \times c + d$

★5. Match the equivalent expressions. Name the property that relates the equivalent expression.

   a. $7 \times f + 12 \times f$

   b. $(7+f) + (12+f)$

   c. $f \times (7+12) \times f$

   i. $(f \times 7 + f \times 12) \times f$

   ii. $(7+12) + (f+f)$

   iii. $f \times (7+12)$

Find the value of each expression.

6. 6.25×0.78

7. $82.7 \div 0.003$

Write each fraction as an equivalent fraction with a denominator that is a power of ten.

8. $\frac{35}{25}$

★9. $\frac{160}{2000}$

★10. Find the prime factorization of 198.

Name _____ Date _____ Period _____

★1. Write each number as a percent.

$\dfrac{100}{2000}$

$\dfrac{100}{200}$

$\dfrac{100}{20}$

★2. Write the percent in decimal notation.

820%

82%

8.2%

0.82%

★3. Max needs 4 batteries. A package of 2 batteries costs $3.75. Tax is 8%.
Write an expression to show how much Max will pay. Find the value.

★4. Find 30% of 240.   **5.** Find 3% of 240.   **6.** Find .3% of 240.

Find the value of the expression.

7. $\dfrac{6}{5} \times \dfrac{5}{6}$ **★8.** $\dfrac{24}{5} \div \dfrac{3}{10}$

Write each decimal as a fraction and as a percent.

9. 2.56   **10.** 0.028

Name _____ Date _____ Period _____

Find the solution to the equation.

1. $8 + m = 42$ ★**2.** $56 = 4 \times s$

What step or sequence of steps turns the equation on the top into the equation on the bottom?

3. $16 = p$ ★**4.** $n = 2$
$18 = p + 2$ $3 \times n + 1 = 7$

Make an equivalent equation as directed.

★**5.** $d = 48$

Step 1: Add 5 to both sides
Step 2: Multiply both sides by 3
Step 3: Evaluate the right side of the equation

Multiply. Express your answer in decimal notation.

6. 10×0.62 **7.** 5.72×10 ★**8.** 7.2×1000

100×0.62 5.72×100 $7.2 \times \frac{1}{100}$

1000×0.62 5.72×1000 $7.2 \times \frac{1}{10}$

Evaluate each expression when $m = 4$, $k = 7$, **and** $d = 8$.

★**9.** $m + k \times (4 + m) + d$

10. $m \times (k + d) + m + k$

Name _____ Date _____ Period _____

Find the length of the missing side.

★**1.**

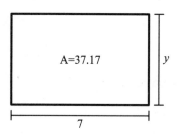

A=37.17

y

7

2.

A=1.44

w

9

Find the value of the variable. Write the value as a fraction.

★**3.** $\frac{2}{7} \times n = 1$

4. $7 \times m = 1$

Use the inverse property of multiplication to solve for the variable.

★**5.** $\frac{1}{9} \times w = 0.78$

5. $\frac{10}{9} \times s = 1.2$

Write an equation to represent the situation. Solve the equation.

7. Two thirds of what number is 17?

★**8.** The sale price of the jacket is 60% of the original price. If the sale price is $57, what was the original price?

Find the value of the expression.

9. $\frac{7}{8} \div \frac{8}{7}$

★**10.** $16 + \frac{4}{3}$

Name _____ Date _____ Period _____

Find the value of the variable by adding the opposite to each side of the equation.

★**1.** $26 + n = 71$

2. $29 = m + (^-3)$

Rewrite subtraction as adding the opposite. Find the value.

★**3.** $38 - 19$

4. $84 - 36$

Represent each situation as an equation. Solve the equation to find the unknown value.

★**5.** 37 less than a number k is 84

6. If h is removed from 72, 18 is left.

Find the multiplicative inverse of each number. Express the value as a fraction.

★**7.** $\frac{7}{12}$

8. 18

★**9.** 64%

10. 25%

Name _____ Date _____ Period _____

Find the value of each expression.

★**1.** $(^-7) + (^-4)$ **2.** $^-7 \times 6$

3. $^-2 + 8$ ★**4.** $^-2 \times 4 \times (^-8)$

Apply the distributive property to the expression.

★**5.** $^-(^-8 + 2)$

Solve each equation.

6. $^-7 + m = 12$

★**7.** $w + (^-6) = ^-2$

8. $k + (^-1) = 1$

Use the inverse property of multiplication to solve for the variable.

★**9.** $\frac{2}{3} \times k = 32.4$

10. $7 \times w = 25.2$

Name _____ Date _____ Period _____

Express the shaded area as a negative rectangle added to a positive rectangle. Evaluate the expression to find the shaded area.

★1.

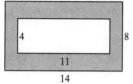

Solve for the variable in each equation.

2. $m + {}^-6 = 6$

★3. $d + ({}^-21) = {}^-8$

Write an equation for each situation.

★4. My friend's meal cost $10 and mine cost $12. We used a coupon which gave us $2 off per person. Our bill before tax and tip was d dollars.

5. This morning there were p sheets of paper in the printer. I printed 10 pages and then refilled the printer with 50 sheets of paper. Now there are 64 sheets of paper in the printer.

Apply the distributive property to each expression.

6. $^-({}^-f + 3)$

7. $^-(9 + {}^-2)$

★8. $^-({}^-6 + {}^-d)$

Find the value of the expression.

9. $^-8 \times {}^-7$

★10. $6 \times (5 + {}^-8)$

Name _____ Date _____ Period _____

Find the value of each expression.

★1. $36 + {}^-87$

2. $^-43 + 81$

★3. $^-42 + {}^-39$

Solve for the variable.

★4. $^-24 = {}^-6 + c$

5. $27 + f + {}^-16 = {}^-5$

★6. The sum of a number and 36 is 17. What is the number?

Find the value of each expression.

★7. $(4+6) \div \frac{1}{10}$

8.
$$
\begin{array}{r}
3\ 2\ 5 \\
\times\ \ \ 4\ 7 \\
\hline
\end{array}
$$

9. $63.7 \div 0.004$

10. $6.25 + 7.1034$

Name _____ Date _____ Period _____

Write an expression for the displacement needed to go from the start point to the end point. Find the value of the expression.

1.

★2.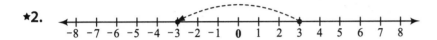

Rewrite subtraction as addition by the opposite. Then find the value of the expression.

3. $7 - 8$

4. $13 - (^-5)$

★5. $^-6 - (^-8)$

Calculate the displacement.

★6. From $^-16$ to $^-48$

7. From $^-17$ to 1

Solve for the variable by adding the same value to both sides of the equation.

★8. $^-5 - (^-12) = g - 7$

Find the rate. Write an equation. Then solve the equation to answer the question.

★9. Mark drove 60 mph for $4\frac{1}{2}$ hours. How far did he drive?

10. 28 beats in 10 seconds is how many beats per minute?

Name _____ Date _____ Period _____

Use the symbol = or ≠ to make the statement true.

★1. $\dfrac{-5}{3} \bigcirc \dfrac{5}{-3}$

2. $\dfrac{-1}{-5} \bigcirc \dfrac{1}{5}$

Define a variable for the unknown value and write an equation to describe the situation. Then solve the equation to answer the question.

★3. Mandy started with $825 in her savings account. She withdrew $25 per week during the summer. How much is left in her account after 12 weeks?

Find the value of the expression. Don't forget the units.

4. $\dfrac{-20 \text{ dollars}}{1 \text{ day}} \times 7 \text{ days}$

★5. $\dfrac{^-4 \text{ inches}}{^-5 \text{ hours}} \times 15 \text{ hours}$

Solve for the variable.

★6. $3 \times m = 24$

7. $t - (^-4) = 18$

Find the value of the expression.

★8. $\dfrac{4}{7} + \dfrac{8}{21}$

9. $\dfrac{1}{9} + \dfrac{13}{12}$

10. $1\dfrac{2}{6} + 3\dfrac{4}{9}$

Algebra Readiness Warm Ups

Name _____ Date _____ Period _____

Place the fraction and it's opposite on the number line.

★**1.**

a. $-\frac{6}{5}$

b. $\frac{1}{3}$

c. $2\frac{7}{8}$

Express each mixed number as a fraction. Show all the steps.

2. $-3\frac{5}{8}$

★**3.** $7\frac{1}{9}$

4. $7\frac{99}{100}$

Find the value of each expression. Write the value as a decimal.

★**5.** $72 \div 9$
$^-72 \div 9$
$72 \div ^-9$
$^-72 \div ^-9$

6. $855 \div 6$
$855 \div ^-6$
$^-855 \div ^-6$
$^-855 \div 6$

Write each fraction as a decimal.

7. $\frac{100}{16}$

★**8.** $\frac{7}{4}$

Find the value of the expression.

9. $\frac{3}{4} + \frac{5}{6}$

★**10.** $3\frac{1}{4} + 2\frac{5}{8}$

Name _____ Date _____ Period _____

★1. Classify each number as a whole number, an integer, or a rational number. Some may belong to more than one category.

   a. $-\dfrac{21}{7}$ **b.** 1

   c. $^{-}9.003$ **d.** $^{-}8$

2. The perimeter is the distance around a rectangle. What is the perimeter of this rectangle?

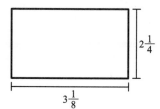

$2\frac{1}{4}$

$3\frac{1}{8}$

Find the multiplicative inverse of each number.

★3. 32%   **★4.** $\dfrac{8}{7}$ **★5.** 1   **6.** $\dfrac{1}{13}$

Solve for the variable in each equation.

7. $6 - k = 4.7$ **8.** $4 \times d = 48$

★9. $t = \dfrac{3}{5} \times \dfrac{1}{9}$ **10.** $7862 \div 0.005 = s$

Name _____ Date _____ Period _____

Fill in the table with solutions to the given equation.

1. $f = \frac{3}{4} \times n$

n	f
-4	
0	
2	
4	

★2. $d = {}^-6g$

g	d
-1	
0	
1	
2	

★3. What scale factor turns a length of 8 into 3?

4. Multiplication by what ratio turns 6 into ⁻10?

Write an equation to describe each situation. Then solve for k.

★5. Twelve less than a number k is $\frac{5}{4}$.

6. Three times a number k is 39.

Find the value of each expression when $g = \frac{2}{3}$.

7. $g + \frac{1}{6}$

★8. $\frac{6}{16} \times g$

Calculate the displacement.

★9. from 31 to ⁻16

10. from $\frac{3}{20}$ to $\frac{1}{4}$

Name _____ Date _____ Period _____

Write an equation to describe each situation. Use the variable d to represent "the number".

1. Half the number is 16.

2. Twice the number is 10 more than 7.

★3. The product of 6 and the number is 54.

★4. Three times the sum of the number and 8 is 33.

State whether or not each expression is equivalent to $4 \times (n + 4) + 2$.

5. $4 \times n + 4 + 2$ ★6. $n \times 4 + 18$

7. $2 + (4 + n) \times 4$

Solve the following equations. Show all the steps.

★8. $n + 7 = {}^-3$ ★9. $2 \times m = 7 + ({}^-11)$

10. $\frac{h}{5} = {}^-10$

Name _____ Date _____ Period _____

1. Name the rule or property to justify each step.

$$5 \times k = 55$$
$$\frac{1}{5} \times 5 \times k = \frac{1}{5} \times 55$$
$$1 \times k = \frac{1}{5} \times 55$$
$$k = \frac{1}{5} \times 55$$
$$k = 11$$

Solve the equation. Show each step.

2. $4x = 15$

3. $9 - p = 15$

4. $\frac{7}{8}y = \frac{1}{100}$

★5. $2 \times k + 1 \times k = 24$

★6. Is $z = 9$ is a solution to the following equation?

$$\frac{^-2}{3} \times z + 6 = {}^-12$$

Find the value of the expression. Remember to include the units.

★7. $\frac{5 \text{ feet}}{3 \text{ min}} \times 12 \text{ min}$

★8. $\frac{\$22}{3 \text{ jars}} \times 12 \text{ jars}$

9. $\frac{353 \text{ words}}{5 \text{ min}} \times 30 \text{ min}$

★10. $\frac{18 \text{ ounces}}{4 \text{ bottles}} \times 0.5 \text{ bottles}$

Name _____ Date _____ Period _____

Solve for the variable in each equation.

1. $\dfrac{\$5}{8 \text{ pens}} \times n \text{ pens} = \$20$

★2. $\dfrac{2 \text{ feet}}{1 \text{ second}} \times \dfrac{60 \text{ seconds}}{1 \text{ minute}} \times \dfrac{60 \text{ minutes}}{1 \text{ hour}} \times 1.5 \text{ hours} = k \text{ feet}$

Solve. Show all work.

★3. A migrating bird traveled 1,000 miles in one day, going an average of 40 mph. How many hours did it fly?

★4. If a heart beats at a constant rate of 18 beats every 10 seconds, how many times will it beat in 3 hours?

5. A lobster costs $0.75 per ounce. How much does a 3 pound lobster cost?
   Note: There are 16 ounces in a pound.

Rewrite each expression using hidden multiplication signs.

★6. $10 \times k - 5 \times j$

7. $6 \times c \times d \times e$

★8. $\dfrac{17}{18} \times (s + t)$

Rewrite each expression to show where the multiplication signs are.

9. $12b$

10. $\dfrac{1}{20}s + \dfrac{3}{20}t + \dfrac{2}{20}u$

Name _____ Date _____ Period _____

Use the given rate to fill in the missing values.

★1.

(input) tables	(output) chairs
2	
4	
10	

→ $\dfrac{\text{3 chairs}}{\text{2 tables}}$ →

2.

(input) minutes	(output) dollars
10	
20	
60	

→ $\dfrac{\text{1 dollar}}{\text{5 minutes}}$ →

★3. Plot each pair of input and output values.

feet	yards
⁻3	0
⁻6	0
9	6

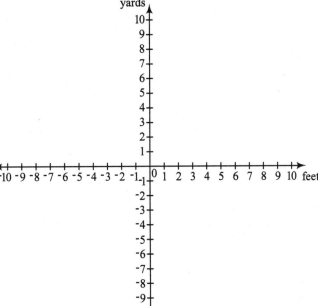

Solve for the variable.

4. $x + 1 = 8$

5. $\dfrac{1}{2}m = 9$

★6. $4 - 3x = 7$

7. $\dfrac{4}{5} \times m = 12$

Use = or ≠ to make each statement true.

★8. $\dfrac{\text{6 pints}}{\text{3 quarts}} \bigcirc \dfrac{\text{10 pints}}{\text{5 quarts}}$

★9. $\dfrac{\text{6 lifeguards}}{\text{4 pools}} \bigcirc \dfrac{\text{21 lifeguards}}{\text{14 pools}}$

10. $\dfrac{\text{5 roses}}{\$12} \bigcirc \dfrac{\text{4 roses}}{\$11}$

Name _____ Date _____ Period _____

Plot each ordered pair.

1. (2, 0)

2. (⁻4, 1)

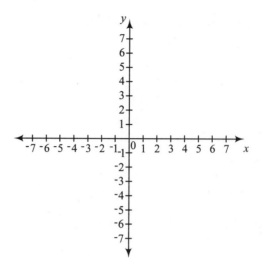

★3. Fill in the table of solutions to the equation.

$$\frac{3}{4}m = n$$

m	n
⁻1	
	0
2	
	9

Write an equation to describe each situation. The input variable is provided. Choose an output variable. Find two solutions to each equation.

★4. Ryan drinks 5 cans of soda every 2 days. At this rate, how many cans of soda will he drink in d days?

★5. It costs $1.50 for every 3 feet of rope. How many feet of rope can you buy for m dollars?

Find the value of the expression.

★6. ⁻6 − 4   ★7. ⁻6 − (⁻4)   8. 3 − (⁻5)   9. $-\frac{3}{4} \times \frac{2}{7}$ 10. $\frac{4}{7} \div \frac{2}{3}$

Name _____ Date _____ Period _____

★1. Plot the points on the coordinate plane.

  a. $(3, {}^-2)$

  b. $({}^-7, 0)$

  c. $(0, 4)$

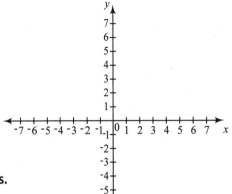

For each equation, fill in the blank in the table of solutions.

2. $\dfrac{2 \text{ calculators}}{3 \text{ students}} \times m \text{ students} = k \text{ calculators}$

m	k
⁻3	
0	
3	
6	

★3. $\dfrac{{}^-5 \text{ points}}{2 \text{ minutes}} \times d \text{ minutes} = y \text{ points}$

d	y
⁻2	
0	
2	
8	

Solve for the variable in each equation.

★4. $n + 7 = 20$

5. ${}^-6 = p - 13$

Find the value of each expression when $m = 6$.

★6. $\frac{2}{3}m + 8$

7. $\frac{1}{2}m + ({}^-6)$

Define a variable, write an equation and solve.

8. The store had a sale where all clearance items were 50% off. If the sale price is $47.38, what was the original price?

9. The machine sews 4 inches every 3 seconds. How many inches can it sew in one hour?

★10. If the phone company charges 45¢ per minute, what will the charges be for 3 hours?

Name _____ Date _____ Period _____

Calculate the displacement.

1. From 0 to 10

★**2.** From 6 to 17

★**3.** From ⁻34 to ⁻67

4. From 125 to ⁻23

Solve for the variable.

5. $\frac{1 \text{ hour}}{60 \text{ miles}} \times 240 \text{ miles} + 1.5 \text{ hours} = m \text{ hours}$

★**6.** $g \text{ gallons} = \frac{15 \text{ gallons}}{2 \text{ cars}} \times 6 \text{ cars}$

Find two solutions to each equation. Plot the solutions on the coordinate plane. Fit a line to the plot to represent the graph of the equation:

★**7.** $g = t + 1$

8. $2w = z$

Identify the rate and offset for each situation. Then write an equation to describe the situation.

★**9.** Each batch of cookies uses 2 pounds of flour. It takes n pounds of flour to make j batches of cookies.

10. A repair shop charges a $25 service fee and then an additional $60 per hour for work done. The charge for a repair that takes t hours is c dollars.

Name _____ Date _____ Period _____

★**1.** Look at the graph of the two runners. Which runner won the race?

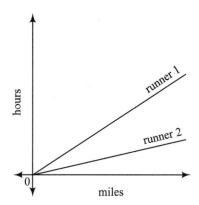

Calculate each slope using the two points given.

2. (1, 6) and (2, 7)

★**3.** (2, 17) and (4, 17)

4. (6, 2) and (7, 5)

★**5.** (⁻3, 3) and (4, 1)

Solve for the variable in each equation.

6. $7 - (^-v) = 18$

★**7.** $6 = 11 + d$

★**8.** $\frac{1}{2}n = {}^-7$

9. $\frac{3}{2} \cdot w - 6 = 6$

10. $3x = 8 - (^-4)$

Name _____ Date _____ Period _____

1. Draw a line that has a negative slope and an offset of 0.

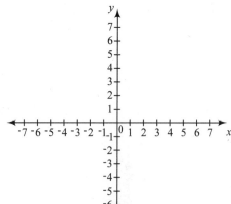

★2. Find the slope of the line passing through (2, 1) and (6, 3).

★3. Solve for k to find the offset.

$$\frac{5 \text{ pounds}}{\$3} \times \$15 + k \text{ pounds} = 20 \text{ pounds}$$

Answer the questions below about the equation.

$$r \text{ cars} \times \left(\frac{10 \text{ clowns}}{1 \text{ car}}\right) + 1 \text{ clown} = c \text{ clowns}$$

★4. Is $r = 2, c = 2$ a solution?

★5. When the input is 5 cars, what is the output?

Solve for the variable, *m*.

6. $3 - (^-2) = m$

7. $^-3 \cdot x + m = y$ if $x = 5$ and $y = 4$

Calculate the displacement.

★8. From 12 to $^-2$

9. From $^-2$ to 0

10. From $^-6$ to $^-1$

Name _____ Date _____ Period _____

★**1.** What scale factor turns a length of 24 into a length of 2?

★**2.** Multiplication by what rate turns $50 into 30 tulips?

★**3.** Start at (0, 3). Make an x change of 2 and a y change of $^-3$. What are the coordinates of the new point?

4. Start at (3, $^-2$). Make an x change of $^-5$ and a y change of $^-3$. What are the coordinates of the new point?

Find the change in x and the change in y from the start point to the end point.

5. Start at ($^-3$, 2) and end at (2, $^-4$)

6. Start at ($^-2$, $^-2$) and end at (6, 0)

Write each fraction as a decimal and as a percent.

★**7.** $\frac{13}{25}$

8. $\frac{22}{5}$

Write each percent as a decimal and as a fraction.

9. 0.4%

★**10.** 476%

Name _____ Date _____ Period _____

Graph each triangle with the given points. Then scale the coordinates by the given factor. Graph the new triangle on the same coordinate plane.

★**1.** (0, 0), (-2,4), (4, 4)

Scale by a factor of $\frac{3}{2}$

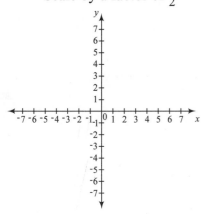

2. (0, 0), (0, 12), (-4,0)

Scale by a factor of 75%

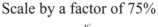
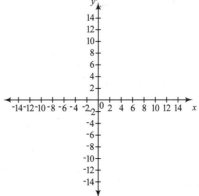

Find the missing coordinate for the similar triangle.

★**3.**

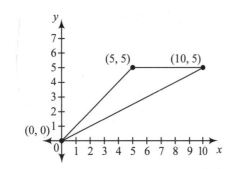

4.

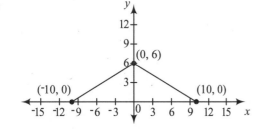
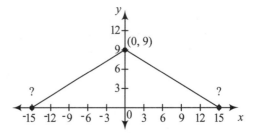

Write each fraction as a percent.

★**5.** $\frac{6}{25}$

★**6.** $\frac{6}{250}$

Multiply or divide.

★**7.** $\frac{5}{8} \cdot 7$

8. $\frac{4}{5} \cdot 19$

9. $75 \div 8$

10. $213 \div 20$

Name _____ Date _____ Period _____

Solve for the variable in each equation. Write the value in decimal notation.

★1. $4s = 6.4$

2. $m \times 0.4 = 82$

Below are pair of similar triangles. Find the missing height.

★3.

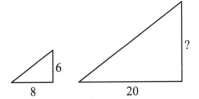

Find the slope of the line that goes through the given points.

★4. (1, 3) and (2, 6)

5. (0, 0) and (-6, 7)

★6. The shadow from a building is 12 feet long. The shadow from a dog is 0.5 feet long. If the dog is 1 foot high, how high is the building?

7. Find 40% of 6.

★8. Find 120% of 50.

9. 25% of what number is 56?

10. Three times what number is 63?

Name _____ Date _____ Period _____

Write an equation for the line that passes through the origin and the given point.

1. $(6, 2)$ ★**2.** $(^-4, 3)$ **3.** $(^-4, 5)$

4. Solve for y when $x = 3$.

$$y = {}^-2x$$

★**5.** Solve for y when $x = \frac{2}{3}$.

$$y = \frac{7}{4}x$$

Simplify each expression.

★**6.** $\dfrac{2\text{ feet}}{5\text{ seconds}} \cdot 30\text{ seconds} + 5\text{ feet}$

7. $\dfrac{35\cancel{c}}{4\text{ oz}} \times \dfrac{16\text{ oz}}{1\text{ pound}} \times 5\text{ pounds}$

8. $\dfrac{50\text{ words}}{6\text{ minutes}} \times \dfrac{60\text{ minutes}}{1\text{ hour}} \times 4\text{ hours} + 85\text{ words}$

★**9.** Are $3x + 6 = 18$ and $x = 4$ equivalent equations?

★**10.** Are $0 \times (a + b)$ and 4×0 equivalent expressions?

Name _____ Date _____ Period _____

Find the value of each expression.

★1. $\dfrac{6 \text{ dollars}}{5 \text{ pounds}} \times 4 \text{ pounds}$

2. $\dfrac{2 \text{ tables}}{11 \text{ people}} \times 77 \text{ people}$

Use = or ≠ to indicate if the ratios are equivalent or not.

★3. $\dfrac{9}{12}$ ◯ $\dfrac{8}{11}$ **4.** $\dfrac{8}{6}$ ◯ $\dfrac{60}{45}$

Determine if the graph of the equation goes through the origin. If it does, find an equivalent equation in $y = mx$ form.

★5. $5x = 7y$ **6.** $y + 1 = 6x - 1$

Below are pairs of similar triangles. Find the missing height.

7.

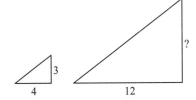

★8.

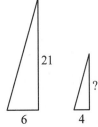

Define a variable to represent the unknown quantity and write an equation to describe the situation. Then solve the equation to answer the question.

9. Six bottles of water cost $2.50. At this rate, how many bottles of water can you buy for $10?

★10. If a train travels at a constant speed of 150 miles every 3 hours, how far will it go in 2 hours?

Name _____ Date _____ Period _____

★1. Find three different fractions that are equivalent to $\frac{3}{4}$.

2. Write the first 10 prime numbers.

Find the prime factorization of each number.

3. 100   4. 16   ★5. 225

Determine if the number is a perfect square or not. If it is, find the square root.

★6. 45   ★7. 81   ★8. 90

Solve for the variable.

9. $n = 12 \times 12$

10. $64 = 8 \times y$

Name _____ Date _____ Period _____

1. Could 3,4, and 6 be the lengths of the sides of a right triangle?

★2. Write the first 10 prime numbers.

Find the prime factorization of each number.

★3. 192

4. 3,600

Indicate if each number is a perfect square or not. If it is, find the square root.

5. 24

★6. 196

Solve for the variable.

7. $9 \times n = 81$

★8. $a^2 + 8^2 = 10^2$

Find the value of each expression.

★9. 13^2

10. 15^2

Name _____ Date _____ Period _____

Find the value of each expression.

1. 3^2

 ★2. 3^3

3. $2^2 \times 5^2$

 ★4. $6 \times 10^3 + 7 \times 10^1 + 4 \times 10^0$

Express the repeated multiplication using exponents.

★5. $13 \times 13 \times 13 \times 13$

6. $7 \times 7 \times 7 \times 7 \times 7 \times 7 \times 7$

Write each number as 10 to a power.

7. One thousand

★8. One million

Solve for the variable.

9. $\frac{1}{3} + \left(-\frac{1}{4}\right) = e$

 ★10. $g \times 100 = 10^3$

Name _____ Date _____ Period _____

Write each expression using exponents.

★**1.** 11

11 × 11

11 × 11 × 11

2. $\frac{1}{10}$

$\frac{1}{10} \times \frac{1}{10}$

$\frac{1}{10} \times \frac{1}{10} \times \frac{1}{10}$

Write each value in expanded form using exponential notation for the powers of 10.

3. 16.4   ★**4.** 907.05

Find the value of each expression. Write your answer in fraction form.

★**5.** 2^{-1} ★**6.** $\left(\frac{4}{3}\right)^2$

7. 15^{-2} **8.** 47^0

Find the value of each expression. Write your answer as a fraction in which the numerator and the denominator do not have any of the same prime factors.

9. $\frac{36}{5} \times \frac{10}{20}$

★**10.** $\frac{1}{16} \div \frac{1}{8}$

Name _____ Date _____ Period _____

Find the value of each expression. Write the value as a decimal.

★**1.** $10^2 \times 10^{-6}$

2. $\left(\frac{5}{3}\right)^{-3}$

Find the reciprocal of each value. Put your answer in fraction form.

★**3.** 0.7

4. 125%

Simplify each expression.

5. $y^4 \times y^0$

★**6.** $\dfrac{n^{-1} \times n^5}{n^{-3}}$

Indicate if the number is a perfect square or not. If it is, find the square root.

★**7.** 80

8. $10{,}000$

Solve for the variable.

9. $d \times 7^{-1} = 1$

★**10.** $1 = \dfrac{10}{12} \times s$

Name _____ Date _____ Period _____

Find the slope of the line that goes through the given points.

★**1.** (2, 3) and (5, 7)

★**2.** (⁻13, 1) and (0, 0)

Which two whole numbers are closest to each value?

3. $\sqrt{17}$

4. $\sqrt{30}$

Solve for x when $y = 0$.

★**5.** $y = \frac{3}{4}x$

★**6.** $y = 2x - 6$

Solve for the variable b in each equation.

★**7.** $\frac{1}{3} \cdot 9 + b = 4$

8. $^{-}5 \cdot 2 + b = 7$

Find the value of each expression.

9. $\sqrt[3]{27}$

10. $\left(\frac{1}{4}\right)^{-1}$

Name _____ Date _____ Period _____

Find the x-intercept and y-intercept for each equation.

★**1.** $-6x = y$

2. $y = \frac{1}{2}x - 3$

Solve for y when $x = 2$.

★**3.** $y = \frac{1}{3}x + \frac{1}{6}$

4. $y = -4x + 3$

Solve for x when $y = 8$.

★**5.** $y = 3x + 7$

6. $y = -x + 6$

Graph each equation on the same coordinate plane.

7. $y = x - 2$

8. $y = 3x + 1$

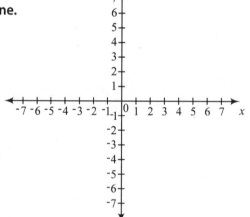

Write an equation to describe each situation. Use y as the output variable.

★**9.** The plant needs $\frac{1}{2}$ cup of water everyday. How much water does it need in x days?

★**10.** I walk my dog 9 times a week. At this rate, how many times will I walk her in x days?

Name _____ Date _____ Period _____

Graph each equation on the same coordinate plane.

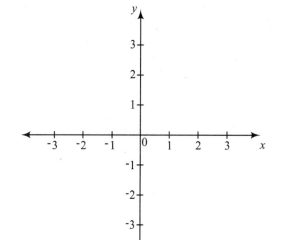

★1. $y = {}^-x$

★2. $y = \frac{3}{2}x + 1$

Solve each system of equations.

★3. $y = 3x$
    $y = x + 4$

4. $y = {}^-4x$
    $y = \frac{2}{5}$

Write an equation to describe each situation. Use the variable n to represent the number.

★5. Half the number is 3 less than 17.

6. The product of the number and 6 is 10 more than 38.

Use <, >, or = to make each statement true.

★7. $\frac{6}{5} \bigcirc \frac{5}{6}$

8. $\frac{9}{9} \bigcirc 9$

9. $\frac{6}{2} \bigcirc \frac{81}{27}$

10. $\frac{11}{30} \bigcirc \frac{12}{31}$

Name _____ Date _____ Period _____

Find the value of each expression.

★**1.** $\dfrac{18 \text{ players}}{4 \text{ referees}} \times 10 \text{ referees}$

2. $\dfrac{1 \text{ minute}}{60 \text{ seconds}} \times 150 \text{ seconds}$

★**3.** $\dfrac{2 \text{ dollars}}{1 \text{ pound}} \times 7.5 \text{ pounds}$

4. $\dfrac{1 \text{ mile}}{10 \text{ minutes}} \times 45 \text{ minutes}$

Write an equation or inequality to describe each situation.

★**5.** The number y is 2 less than half of the number x.

6. The number y is at least 6 more than the number x.

★**7.** The number y is 10 less than the product of x and 7.

Define a variable and write an equation to describe the situation. Then solve the equation to answer the question.

★**8.** Each box weighs 7 pounds. How much do 8 boxes weigh?

9. A shirt is on sale for 20% off the original price. The sale price is $19.20. What was the original price?

10. A florist sells 12 tulips for $18. How much will he charge for 4 tulips? (Assume that the price per tulip is the same regardless of how many are purchased).

Name _____ Date _____ Period _____

Find the value of each expression.

★1. $364 \div 7$

2. $225 \div 9$

★3. $\frac{1}{8} \times 9{,}088$

★4. $2{,}997 \div 3$

Write each decimal as a fraction.

★5. 0.81

6. 1.36

Solve for the variable in each equation.

7. $a = 0.5 \times 0.9$

★8. $6 \times r = 72$

Find the value of each expression. Write the value as a fraction.

9. $\left(\frac{1}{3}\right)^{-2}$

10. $6^2 \times 10^{-2} \times \frac{10^5}{2^{-1}}$

Name _____ Date _____ Period _____

Find the value of each expression. Put your answer in decimal form.

1. $7,635 \times 10$

   $7,635 \times 100$

   $7,635 \times 1,000$

★2. $7,635 \times \frac{1}{10}$

   $7,635 \times \frac{1}{100}$

   $7,635 \times \frac{1}{1,000}$

Find the prime factorization of each number.

★3. 100 ★4. $1,000$

Write each fraction as a decimal. Use bar notation for repeating decimals.

5. $\frac{10}{20}$ 6. $\frac{10}{40}$

★7. $\frac{10}{60}$ ★8. $\frac{10}{80}$

Find the value of the variable.

9. $10.76512 + (^-8.06512) = d$

10. $y + \frac{2}{7} = 8\frac{5}{7}$

Algebra Readiness Warm Ups

Name _____ Date _____ Period _____

Solve for the variable in each equation.

★**1.** $\frac{3}{2}y = 75$

★**2.** $10 = k + 16$

★**3.** $^{-}d = ^{-}5$

4. $2 = z \times 17$

★**5.** Is $n = 3$ is a solution to the equation $4n = 3n + 2$?

6. Is $w = 26$ is a solution to the equation $214 = 8 \times w + 6$?

★**7.** Write an equation that is equivalent to $12 \times (t + 1) = 55$.

8. Express each number in decimal notation. Use bar notation for repeating decimals.

   a. $\frac{1}{5}$

   b. $\frac{1}{6}$

   c. $\frac{1}{9}$

Without doing any division, state whether each fraction will be a repeating decimal or a terminating decimal and justify your answer.

9. $\frac{3}{8}$

10. $\frac{8}{3}$

Algebra Readiness Warm Ups

Name _____ Date _____ Period _____

Evaluate each expression when $s = 10$.

★**1.** $\frac{1}{5}(20-s)$ **2.** $\frac{1}{5}s + \frac{2}{3}$

Find the variable x in each equation.

★**3.** $6x + 2 = 14$

4. $5x - 75 = 0$

5. $2(x-3) = 48$

6. $10 = \frac{1}{3}x$

Represent each situation with an equation. Use the variable y to represent the number.

7. Five more than twice the number is 17.

★**8.** Half of the number is 10 less than 60.

Define a variable for the unknown quantity. Write an equation to represent the situation. Solve the equation to answer the question.

★**9.** 16% of what number equals 44?

★**10.** Dmitry bought a shirt for 40% of the regular price. If he paid $36.42, what was the regular price?

Name _____ Date _____ Period _____

Use the distributive property to write an equivalent expression.

★1. $(3+7)x$

2. $ab + ac$

State whether or not each equation is equivalent to $3(2+x) + 7 = 28$.

★3. $6 + x + 7 = 28$ **4.** $x = 5$

★5. $21 = 3(x+2)$ **6.** $3(2+x) + 1 = 4$

Solve each equation.

★7. $9 = 5 - 2x$

8. $\frac{4}{5} - b = \frac{3}{5}$

Define a variable, write an equation and solve.

★9. The final exam is 10 pages longer than the midterm. The final is 18 pages. How many pages is the midterm.

10. One large pizza is $3 less than two small pizzas. If a large pizza costs $15, how much does a small pizza cost?

Name _____ Date _____ Period _____

Write each expression as 3 raised to a power.

★**1.** $3 \times 3 \times 3 \times 3$

2. $\frac{1}{3}$

★**3.** $\frac{1}{9}$

4. 1

Find the value of each expression. Write the value in decimal notation.

★**5.** $(10^3)(10^{-7})(10^2)$

6. $(2^2)^{-1}$

Solve each equation.

★**7.** $\frac{11}{3}u = 1$

8. $q^{-1} = \frac{1}{7}$

Simplify and then solve.

★**9.** $x + x + 2x = 13$

10. $\frac{1}{6}x - 3 + \frac{17}{6} = 7$

Name _____ Date _____ Period _____

Find the reciprocal of each number. Write the result as a fraction.

★**1.** 7

2. $\frac{1}{30}$

3. 1.2

★**4.** 64%

Rewrite each expression without using any negative exponents.

5. $x^{-1}y^{-1}$

★**6.** $\left(\frac{t^{-1}}{u^{-1}v}\right)^{-1}$

Find the value of each expression. Write the value in decimal notation.

★**7.** $\left(\frac{2}{3}\right)^{-1}$

8. $\frac{\frac{1}{2}}{\frac{5}{7}}$

Solve for the variable x in each equation.

9. $\frac{3^{-1}}{4^{-1}} = x + \frac{2}{3}$

★**10.** $\left(\frac{7}{11}\right)^{-1}x = 11$

ANSWERS

Lesson 3 Warm Up

1. $m = 16$

2. $p = 17$

3. $v = 15$

4. Expression

5. Expression

6. Equation

7. Expression

8. a is negative

   b is negative

   c is positive

   d is positive

   e is negative

9. 2

10. 7

Lesson 4 Warm Up

1. **a.** 10   **b.** 15   **c.** 3

2. **a.** 15   **b.** 4   **c.** 13

3.

f	h
0	3
4	7
5	8
7	10

4. No

5. Yes

6. Yes

7. $m = 8$

8. $n = 1$

9. $k = 6$

10. $f = 9$

Lesson 5 Warm Up

1. $v + 12$

2. $c + d + 18$

3. **a.** 17   **b.** 13   **c.** 11

4.

e	g
1	4
2	5
6	9
5	8

5. **a.** 8   **b.** 9   **c.** 41

6. $m = 7$

7. $d = 4$

8. $f = 2$

9. $y = 7$

10. $m = 6$

Lesson 6 Warm Up

1. 7×6

2. $3 \times n$

3. **a.** 42   **b.** 54   **c.** 24

4. $4 \times (c + 3)$

5. $3 \times (d + f + 1)$

6. $m = 0$

7. $n = 5$

8. $d = 1$

9. $n = 3$

10. $w = 4$

10.

Lesson 7 Warm Up

1. $6 \times 4 = 24$

2. $9 \times 1 = 9$

3. $3 \times 8 = 24$

4. $4 + 2 + 6 = 12$

5. 24 units long

6. $t = 4$

7. $m = 3$

8. $w = 9$

9. $y = 8$

10. $c = 11$

Lesson 8 Warm Up

1. $3 \times 5 + 4 \times 3 = 27$

2. $4 \times 1 + 7 \times 4 + 4 \times 2 = 40$

3. 40

4. 35

5. 54

6. $t = 56$

7. $n = 14$

8. $s = 1$

9.

Lesson 9 Warm Up

1. n is Mario's points; $n = 10 + 7$

2. n is how much money I have now; $n + 3 = 22$

3. n is how many hamburgers we bought; $n \times 2 = 12$

4. n is how many hours Melissa worked on Saturday; $n \times 9 = 36$

5. n is how many slices of pizza I ate this afternoon; $3 + 2 + 1 + n = 8$

6. $3 \times (e + 3)$

7. $3 \times (f + g + 2)$

8. $6 \times (4 + 3)$

9. $1 + 2 + 3 + 4 + 5 + 6$

10. n is how many tables there are; $n \times 6 = 42$; $n = 7$

Lesson 10 Warm Up

1. $3 \times (f + g)$

2. $5 \times (7 + d + e)$

3. $k \times (2 + m + p)$

4. a. iii **b.** iv **c.** ii **d.** i

5. $a = 5$; Distributive property

6. $a = 0$

7. $c = 18$

8. $d = 0$

9. $f = 1$

10. $g = 5$

ANSWERS

Lesson 11 Warm Up

1. $8 \times 100 + 6 \times 10 + 5 \times 1$

2. $4 \times 1{,}000 + 6 \times 100 + 2 \times 10 + 3 \times 1$

3. $3 \times 10 + 5 \times 1$

4. $2 \times 1{,}000 + 4 \times 10 + 1$

5. $2 \times 100 + 4 \times 10$

6. $1 \times 1000 + 3 \times 100 + 3 \times 10 + 5 \times 1$

7. $9 \times 100 + 2 \times 10 + 5 \times 1$

8. $7 \times (g + h + 3)$

9. $m \times (3 + p)$

10. $d = 100$

Lesson 12 Warm Up

1. 6,704,649

2. 80,602,005,080

3. $6 \times 1{,}000{,}000{,}000 + 3 \times 100{,}000{,}000$
$+ 5 \times 10{,}000{,}000 + 7 \times 100{,}000 + 3 \times 1$

4. 70,004,000,000,559

5. Six billion, two hundred fifty-seven million, three hundred four thousand, eight hundred sixty-one.

6. $8 \times (4 + 3)$

7. $m \times (6 + 3 + 2)$

8. $10 \times (5 + 3)$

9. $100 \times (3 + 2 + 1)$

10. $1{,}000 \times (k + p)$

Lesson 13 Warm Up

1. $6 \times 10^2 + 1 \times 10^1 + 2 \times 10^0$

2. $4 \times 10^{11} + 3 \times 10^8 + 3 \times 10^6 + 9 \times 10^5 + 6 \times 10^2 + 7 \times 10^1 + 5 \times 10^0$

3. 3,700,106,804

4. $3 \times 10{,}000{,}000 + 8 \times 100{,}000$

5. $1 \times 1{,}000{,}000{,}000 + 3 \times 100{,}000{,}000 + 9 \times 1{,}000 + 1 \times 100$

6. 10^0

7. 10^3

8. 10^5

9. 152

10. 28,138

Lesson 14 Warm Up

1. $4 \times 100 + 2 \times 1$

2. $5 \times 100 + 5 \times 10 + 2 \times 1$

3. $2 \times 7 \times 10 + 2 \times 3 \times 1$;
Expanded form: $1 \times 100 + 4 \times 10 + 6 \times 1$

4. $6 \times 4 \times 10 + 6 \times 8 \times 1$;
Expanded form: $2 \times 100 + 8 \times 10 + 8 \times 1$

5. $182

6. $14

7. 231

8. 3,616

9. 56,745

10. **a.** 10×10 **b.** $10 \times 10 \times 10 \times 10 \times 10$

Lesson 15 Warm Up

1. 75,000 meters

2. 1,000,000 meters

3. 1,500; 15,000; 150,000; 1,500,000

104 | Answers to Lesson Warm Ups

© Copyright MIND Institute   Algebra Readiness Warm Ups

4. 56; 560; 56,000; 56,000,000

5. n is how many feet the football team runs;
$n = 10 \times (360 + 160 + 360 + 160)$
$n = 10,400$

```
        +160 feet
      ┌──────────┐
      │          │
+360  │          │  +360
feet  │          │  feet
      │          │
      └──────────┘
        +160 feet
```

6. 10^8

7. 10^0

8. $n = 1,000$

9. $d = 10$

10. $y = 9,657$

Lesson 16 Warm Up

1. $\dfrac{1 \text{ dollar}}{20 \text{ nickels}}$

2. $\dfrac{2 \text{ pounds}}{3 \text{ dollars}}$

3. From days to hours; $\dfrac{24 \text{ hours}}{1 \text{ day}}$; There are 168 hours in 7 days.

4. From dollars to dividers; $\dfrac{5 \text{ dividers}}{2 \text{ dollars}}$; You can buy 25 dividers for $10.

5. From kilometers to meters; $\dfrac{1,000 \text{ meters}}{1 \text{ kilometer}}$; There are 10,000 meters in 10 km.

6. 703,287

7. Six hundred twenty-five million, five hundred three

8. 2,115

9. 2,664

10. 2,980

Lesson 17 Warm Up

1. $\dfrac{1 \text{ hour}}{60 \text{ minutes}}$

2. $\dfrac{10 \text{ pennies}}{1 \text{ dime}}$

3. $\dfrac{1 \text{ student that walks to school}}{4 \text{ students}}$; If there are 28 students in the class, 7 students walk to school.

4. $\dfrac{3 \text{ dollars}}{4 \text{ oranges}}$; A dozen oranges cost $9.

5. $\dfrac{6 \text{ successful shots}}{10 \text{ shots}}$; If he shoots 30 shots, he will make 18 shots.

6. $10 \times 10 \times 10 \times 10 \times 10$

7. $m + 4 + m + 4 + m + 4$

8. 400,358

9. 81,186

10. 49,218

Lesson 18 Warm Up

1. $\dfrac{5}{6}$

2. 14

3. 30

4. 2

5. $k = 7$

6. $t = 6$

7. 9 fods

8. 32 jags

9. $\dfrac{10 \text{ seconds}}{3 \text{ multiplication facts}}$; It takes 50 seconds to do 15 multiplication facts.

10. $\dfrac{4 \text{ hair clips}}{7 \text{ dollars}}$; 24 hair clips can be bought for $42.

ANSWERS

Lesson 19 Warm Up

1. 18

2. 54

3. 56

4. $\frac{7}{4} \times (12 + 8) = 35$

5. $\frac{1}{5} \times (7 + 8) = 3$

6. $k = 2$

7. $k = 15$

8. 119,063

9. 870,000

10. 5,701

Lesson 20 Warm Up

1. 36 feet

2. $48

3. $\frac{5 \text{ faces}}{2 \text{ pages}}$

4. n is how many pounds of bananas can be bought for $6.
$n = \frac{3 \text{ pounds of bananas}}{\$2} \times \$6$
$n = 9$ pounds of bananas

5. n is how many gallons of gas needed to go 90 miles.
$n = \frac{1 \text{ gallon}}{9 \text{ miles}} \times 90$ miles
$n = 10$ gallons

6. $a = 224$

7. $c = 9$

8. $f = 2,740$

9. $n = 7$

10. $m = 9$

Lesson 21 Warm Up

1. $63 \times \frac{5}{7} = 45$

2. 48 minutes $\times \frac{7 \text{ pages}}{6 \text{ minutes}} = 56$ pages

3. $63 \div 7 = 9$

4. $\$24 \div \frac{\$8}{1 \text{ hour}} = 3$ hours

5. $a = 5$

6. $m = 8$

7. $t = 15$

8. $d = 16$

9. $f = \frac{3}{2}$

10. $w = 0$

Lesson 22 Warm Up

1. $63 \div 9$; 7

2. 5×144; 720

3. 28; 1, 2, 4, 7, 14, 28

4. 1, 2, 3, 5, 6, 10, 15, 30

5. 1, 19

6. $72 \times \frac{7}{9}$; 56

7. $48 \times \frac{7}{6}$; 56

8. Yes

9. No

10. Yes

Lesson 23 Warm Up

1. $r = 2$

2. $r = 1$

3. 4; 400; 40,000

4. $g = 2$

5. $a = 3; b = 30; c = 300$

6. 22,925

7. 293

8. 63 marbles $\times \dfrac{4 \text{ players}}{9 \text{ marbles}}$; 28 players

9. 36 monkeys $\times \dfrac{5 \text{ acres}}{9 \text{ monkeys}}$; 20 acres

10. 24 feet $\times \dfrac{3 \text{ years}}{2 \text{ feet}}$; 36 years

Lesson 24 Warm Up

1. 465

2. 3,462

3. $c = 912$

4. $d = 417$

5. n is the difference between the heights of redwoods and saguaro cactus; 53 feet $+ n = 367$ feet; $n = 314$ feet

6. $84 = 4 \times 21 + 0$

7. $72 = 3 \times 24 + 0$

8. $61 = 5 \times 12 + 1$

9. $52 = 6 \times 8 + 4$

10. $85 = 7 \times 12 + 1$

Lesson 25 Warm Up

1. $4601 \div 5$

2. $6\overline{)84}$

3. $9\overline{)5}$

4. 126

5. 1389

6. $6 \times \dfrac{12}{6}$; 12

7. $48 \times \dfrac{8}{6}$; 64

8. $24 \times \dfrac{3}{4}$; 18

9. 6 tickets

10. $105

Lesson 26 Warm Up

1. 219 books

2. 138 quarts

3. n is how many fewer kilometers to travel if I leave from Los Angeles; 3,943 km $+ n = 4,139$ km; $n = 196$ km

4. n is the average speed for the round trip; $n = \dfrac{7 \text{ miles} + 7 \text{ miles}}{23 \text{ minutes} + 12 \text{ minutes}}$; $n = \dfrac{2 \text{ miles}}{5 \text{ minutes}}$

5. n is how many cartons of chocolate milk the cafeteria will receive if they order 174 regular milk cartons; $n = 174 \times \dfrac{2}{3}$; $n = 116$ cartons

6. 5×1

7. 6×6

8. $k = 1,084$

9. $m = 942$

10. $y = 167$

Lesson 27 Warm Up

1. $\dfrac{7}{5}$

2. $\dfrac{6}{3}$ or $\dfrac{2}{1}$

3. $3 \times \dfrac{1}{5}$

4. $18 \times \dfrac{1}{4}$

5. $4 > \dfrac{4}{4}$

6. 4

7. 42 jobs

8. 9 units

9. 24

10. 28

Lesson 28 Warm Up

1. $a = 15$

2. $b = 5$

3. $60

4. $n = 621 \times \frac{5}{9}$; $n = 345$ students

5. a. m is how much Monique will save;
   $m = \$84 \times \frac{1}{3}$; $m = \$28$

   b. n is the sale price of the dress;
   $n + \$28 = \$84$; $n = \$56$

6. $\frac{3}{8} \times (20 + 4)$; 9

7. $\frac{1}{3}(8 + 7)$; 5

8. $8 \times 1{,}000 + 7 \times 10 + 8 \times 1$

9. $j = 1{,}007$

10. $z = 3{,}248$

Lesson 29 Warm Up

1. 8×38

2. $\frac{4}{5}$

3. $\frac{16}{3}$

4. $6 \times \frac{1}{4}$

5. $36 \times \frac{1}{8}$

6. $n = 6$

7. $y = 4$

8. $m = 10$

9. $d = 5$

10. $m = 64$

Lesson 30 Warm Up

1. $\frac{3}{8} \times \frac{3}{4}$

2. $\frac{1}{2} \times \frac{3}{6}$

3. $\frac{21}{35} = \frac{3}{5}$

4. $\frac{56}{3}$

5. $\frac{2}{72} = \frac{1}{36}$

6. $a = 8$

7. $m = 9$

8. 5,400,208

9. 72,271

10. $\frac{30 \text{ problems}}{40 \text{ minutes}} = \frac{3 \text{ problems}}{4 \text{ minutes}}$

Lesson 31 Warm Up

1. $\frac{4}{6}, \frac{6}{9}, \cdots$

2. $\frac{2}{10}, \frac{3}{15}, \cdots$

3. $\frac{9}{24}, \frac{18}{48}, \frac{24}{64}$

4. $\frac{28}{24}, \frac{35}{30}, \frac{63}{54}$

5. n is the fraction of the pizza Mark ate;
   $n = 2 \times \frac{1}{3} \times \frac{1}{4}$; $n = \frac{2}{12}$

6. a is the area of the whole quilt in square inches;
   $a = 80 \times 8$; $a = 640$

7. $\frac{4}{7}$

8. 292,218

9. 30

10. 47

Lesson 32 Warm Up

1. $\frac{7}{5} \times \frac{4}{3}$

2. $\frac{24}{35}$

3. $\frac{63}{50}$

4. $\frac{5}{12} > \frac{3}{8}$

5. $\frac{4}{6} = \frac{6}{9}$

6. $\frac{24}{25}$

7. $\frac{42}{40}$

8. $\frac{65}{168}$

9. 425 apples

10. 153 hours

Lesson 33 Warm Up

1. 329 dollars

2. $\frac{7 \text{ dollars}}{2 \text{ pounds}} = \frac{329 \text{ dollars}}{94 \text{ pounds}}$

3. $\frac{6 \text{ days}}{4 \text{ hours}} = \frac{15 \text{ days}}{10 \text{ hours}}$

4. $\frac{40 \text{ dollars}}{\text{hour}}$

5. $\frac{1600 \text{ dollars}}{\text{week}}$

6. Yes

7. No

8. No

9. 1

10. $n = 7$

Lesson 34 Warm Up

1. Even: 24, 38, 254, 150; Odd: 57, 89

2. 66 ÷ 4 has a remainder of 2, so 66 is not divisible

by 4

3. 7, 14, 21, 28, 35

4. Groups of 4: yes; groups of 5: no; groups of 6: yes; groups of 8: no

5. No, because 86 is not divisible by 4 or by 5.

6. 604

7. 8

8. 2892

9. $\frac{5}{6}$

10. $\frac{16}{15}$

Lesson 35 Warm Up

1. 2 – prime, 14 – composite, 9 – composite

2. 17 – prime, 23 – prime, 49 – composite

3. 2, 3, 5, 7, 11, 13, 17, 19, 23, 29

4. 0, 9, 18, 27, 36

5. 2,186 ÷ 3 has a remainder of 2, so 2,186 is not divisible by 3

6. 1, 2, 3, 4, 6, 9, 12, 18, 36

7. 1,652

8. $\frac{85}{24}$

9. 38,520

10. $\frac{35}{342}$

Lesson 36 Warm Up

1. **a.** No **b.** Yes **c.** Yes

2. **a.** Yes **b.** No **c.** Yes

3. **a.** No **b.** No **c.** Yes

4. **a.** Yes **b.** Yes **c.** No

5. 512,604

6. $5 \times 10^2 + 1 \times 10^1 + 2 \times 10^0$

7. $7 \times 10^3 + 2 \times 10^2 + 6 \times 10^1 + 3 \times 10^0$

8. 2, 3, 5, 7, 11, 13, 17, 19, 23, 29

9. 7 – prime, 9 – composite, 27 – composite, 39 – composite

10. 2 + 3 + 6 = 11 so 236 is not divisible by 3, which is a factor of 6, so 236 is not divisible by 6

Lesson 37 Warm Up

1. 1 – neither, 2 – prime, 9 – composite, 21 – composite, 23 – prime, 39 – composite, 43 – prime, 59 – prime, 87 – composite

2. 2, 3, 5, 7, 11, 13, 17, 19, 23, 29

3. $2 \times 2 \times 3 \times 5$; 60

4. $2 \times 2 \times 2 \times 2 \times 3$

5. $2 \times 3 \times 3 \times 13$

6. $2 \times 2 \times 5 \times 5$

7. $2 \times 2 \times 2 \times 5 \times 5 \times 5$

8. $n = 10$

9. $n = 100$

10. $n = 100$

Lesson 38 Warm Up

1. 520

2. 82,000

3. 712

4. $2 \times 2 \times 2 \times 2 \times 2 \times 5 \times 5$

5. $2 \times 3 \times 3 \times 5 \times 5$

6. $2 \times 2 \times 2 \times 2 \times 2 \times 2 \times 2$

7. $1 \times 10^4 + 2 \times 10^1 + 9 \times 10^0$

8. $4 \times 10^7 + 3 \times 10^6 + 8 \times 10^4 + 7 \times 10^3 + 2 \times 10^1 + 9 \times 10^0$

9. $135

10. 192

Lesson 39 Warm Up

1. $2^3 \times 11$

2. $2^3 \times 3 \times 5$

3. 15

4. Yes

5. No

6. $3 \times 3 \times 5 \times 7 \times 7 \times 7$

7. $11.52

8. 25,000 sheets

9. $\frac{3}{2} \times (5 + 3)$; 12

10. $\frac{5}{3} \times (5 + 10)$; 25

Lesson 40 Warm Up

1. $\frac{2 \times 2 \times 2 \times 3}{2 \times 3 \times 3}$; $\frac{4}{3}$

2. $\frac{2 \times 3 \times 5}{2 \times 2 \times 2 \times 2 \times 2}$; $\frac{15}{16}$

3. $\frac{2 \times 2 \times 3}{2 \times 3 \times 3 \times 5}$; $\frac{2}{15}$

4. $2.50

5. 15 cents

6. $\frac{6}{5}$

7. $\frac{10}{27}$

8. $\frac{14}{3}$

9. $\frac{2}{9}$

10. $\frac{32}{25}$

Lesson 41 Warm Up

1. $\frac{7}{2 \times 5}$ and $\frac{5}{2 \times 2 \times 2}$; $\frac{28}{40}$ and $\frac{25}{40}$

2. $\frac{4}{3 \times 3}$ and $\frac{1}{2 \times 2 \times 3}$; $\frac{16}{36}$ and $\frac{3}{36}$

3. $\frac{32}{60} < \frac{35}{60}$

4. $\frac{30}{18} > \frac{24}{18}$

5. The store that sells 18 cans for $7

6. 131 R3

7. 400 R7

8. 14 R2

9. 271 R2

10. 1,450 R0

Lesson 42 Warm Up

1. $\frac{13}{9}$

2. $\frac{13}{5}$

3. $\frac{23}{12}$

4. $6\frac{5}{17}$

5. $9\frac{6}{7}$

6. $\frac{11}{12}$

7. $\frac{7}{12}$

8. $\frac{25}{42}$

9. $\frac{5}{2}$

10. $\frac{45}{4}$

Lesson 43 Warm Up

1. $1\frac{7}{8}$

2. $5\frac{8}{9}$

3. $\frac{5}{2}$

4. $\frac{70}{8}$

5. $\frac{32}{5}$

6. $1\frac{1}{2}$

7. $\frac{5}{28}$

8. $\frac{17}{6}$

9. $\frac{10}{21}$

10. $\frac{25}{24}$

Lesson 44 Warm Up

1. $f = \frac{6}{8}$

2. $k = \frac{1}{4}$

3. $m = \frac{21}{8}$

4. $v = 2\frac{5}{6}$

5. $\frac{1}{15}$

6. 12

7. Two dollars and a quarter

8. $5 \times 10^5 + 2 \times 10^4 + 1 \times 10^3 + 3 \times 10^0$

9. $\frac{4}{5}$

10. $\frac{7}{10}$

Lesson 45 Warm Up

1. $\frac{85}{10}$

2. $\frac{16}{10}$

3. $\frac{80}{10}$

4. $\frac{7}{10}$

5. $\frac{15}{10}$

6. $c = \frac{25}{8}$

7. $m = 8\frac{1}{15}$

8. $w = \frac{56}{5}$

9. $y = 2\frac{1}{2}$

10. $a = \frac{31}{6}$

Lesson 46 Warm Up

1. $\frac{350}{100}$

2. $\frac{56}{100}$

3. 85%

4. 200%

5. $\frac{76}{100} = \frac{19}{25}$

6. $\frac{150}{100} = \frac{3}{2}$

7. $2\frac{1}{4}$ meters

8. 120 centimeters

9. $2^3 \times 5^3$

10. $2^3 \times 5^4$

Lesson 47 Warm Up

1. $\frac{50}{1000}$

2. $\frac{168}{1000}$

3. $h = \frac{876}{1000}$

4. $n = \frac{45}{100} = \frac{9}{20}$

5. 71%

6. 67%

7. $\frac{32}{100}; \frac{8}{25}$

8. $\frac{200}{100}; 2$

9. 15%

10. 4%

Lesson 48 Warm Up

1. $\frac{3}{8}$; equivalent decimal fraction is $\frac{375}{1000}$

2. $\frac{7}{4}$; equivalent decimal fraction is $\frac{175}{100}$

3. $\frac{7}{10} + \frac{6}{100}$

4. $\frac{2}{10} + \frac{4}{100} + \frac{6}{1000}$

5. $\frac{8}{10} + \frac{1}{100}$

6. $2 + \frac{3}{10} + \frac{6}{100}$

7. $\frac{243}{500}$

8. $\frac{52}{125}$

9. $\frac{165}{1000}$

10. $\frac{225}{100}$

Lesson 49 Warm Up

1. 0.509

2. 603.07

3. 0.039

4. 0.07

5. $\frac{536}{10000}$; 0.0536

6. $\frac{168}{100}$; 1.68

7. 0.75; 75%

8. 1.9; 190%

9. $4 \times 10 + \frac{1}{10} + \frac{4}{1000}$

10. $\frac{5}{10} + \frac{2}{100}$

Lesson 50 Warm Up

1. $3 \times 1 + \frac{2}{100}$

2. $\frac{9}{10} + \frac{8}{100}$

3. $\frac{3}{1000}$

4. $\frac{16}{10}$; $\frac{8}{5}$

5. 708.062

6. Three thousand two hundred one and four hundredths

7. $\frac{6}{10}$; 60%

8. 4.0; 400%

9. $6 \times m + 6 \times 4$

10. $\frac{2}{5} \times (3 + 4)$

Lesson 51 Warm Up

1. 17.26

2. 17.649

3. $p = 13.12$

4. $b = 0.61$

5. $x = 9.05$

6. $m = 7.625$

7. The rest of the counter is 1.97 meters long.

8. 43,806

9. $\frac{55}{100} = \frac{11}{20}$

10. 6,000

Lesson 52 Warm Up

1. 0.2

2. 0.0216

3. $\frac{5}{100}$

4. $\frac{326}{100}$

5. $(\$3.96 + \$0.65 + \$2.59) \times 0.1 = \$0.72$

6. $\$76.50 \times 0.08 = \$6.12$

7. $\frac{2}{3} \times (7 + 5)$

8. $7 \times \frac{1}{10} + 7 \times \frac{1}{100}$

9. 5%

10. 4%

Lesson 53 Warm Up

1. 1,555

2. 46.78

3. 0.3286

4. 6.232 cm²

5. 3.114 m²

6. $\frac{3}{100}$

7. $\frac{127}{100}$

8. $9.26

9. $3.79

10. 24 centimeters

Lesson 54 Warm Up

1. $\frac{3.5 \text{ dollars}}{1 \text{ hamburger}}$

2. $\frac{3.6 \text{ words}}{1 \text{ inch}}$

3. 43.434 centimeters

4. 15.24 centimeters

5. $\frac{8}{15}$

6. $\frac{191}{45} = 4\frac{11}{45}$

7. 8

8. 2,688

9. 7,602

10. 1,098 feet

ANSWERS

Lesson 55 Warm Up

1. 3.26

2. 9.744

3. $k = 19.575$

4. $m = 130.6$

5. $\dfrac{1.595 \text{ meters}}{1 \text{ minute}}$

6. 0.85 and 85%

7. 0.76 and 76%

8. 1.8 and 180%

9. 1.08 and 108%

10. 0.75 and 75%

Lesson 56 Warm Up

1. $0.8\overline{3}$

2. 5.6

3. 0.875

4. 1.666667

5. 4.031431

6. 1 is a repeating decimal, 2 is a terminating decimal, 3 is a terminating decimal, 4 is a repeating decimal, 5 is a repeating decimal.

7. Terminating decimal because the only prime factor in the denominator is 2.

8. Repeating decimal because the fraction is simplified and the denominator has a factor of 3.

9. 0.112

10. 2.21

Lesson 57 Warm Up

1. 32.6; 326; 3260

2. 4.09; 40.9; 409

3. $\dfrac{1{,}710.3}{0.8} = \dfrac{17{,}103}{8}$

4. $\dfrac{4{,}230}{8} = 528.75$

5. $\dfrac{714{,}000}{6} = 119{,}000$

6. $\dfrac{30}{5} = 6$

7. 1.25; 125%

8. 0.65; 65%

9. $v = 7$

10. $m = 12$

Lesson 58 Warm Up

1. $\dfrac{1{,}470}{5} = 294$

2. $\dfrac{1{,}728}{3} = 576$

3. $\dfrac{3}{4} = 0.75$

4. $\dfrac{1}{10} = 0.1$

5. $13 + 6.25 + 0 = 19.25$

6. **c.** distributive property

7. **d.** commutative property of addition

8. **b.** associative property of multiplication

9. **e.** multiplicative property of 0

10. **a.** commutative property of multiplication

Lesson 59 Warm Up

1. $7 \times 32 + \frac{7}{4} \times 6$

2. $(4+5) \times (m+8)$

3. $(5 \times 0.6 \times p) + (3 \times f) + (f)$

4. $(a \times b) + (6 \times c) + (d)$

5. a. iii - distributive property

   b. ii. - commutative property of addition

   c. i. - distributive property

6. 4.875

7. $\frac{82,700}{3}$ or $27,566.\overline{6}$

8. $\frac{14}{10}$

9. $\frac{80}{1000}$ or $\frac{8}{100}$

10. $2 \times 3 \times 3 \times 11 = 2 \times 3^2 \times 11$

Lesson 60 Warm Up

1. $\frac{100}{2000} = 5\%$, $\frac{100}{200} = 50\%$, $\frac{100}{20} = 500\%$

2. $820\% = 8.2$; $82\% = 0.82$; $8.2\% = 0.082$; $0.82\% = 0.0082$

3. $x = (2 \times 3.75 \times 0.08) + (2 \times 3.75) = \$8.10$

4. 72

5. 7.2

6. 0.72

7. 1

8. 16

9. $\frac{256}{100}$; 256%

10. $\frac{28}{1000}$; 2.8%

Lesson 61 Warm Up

1. $m = 34$

2. $s = 14$

3. $16 + 2 = p + 2$

4. $3 \times n = 3 \times 2$

   $3 \times n = 6$

   $3 \times n + 1 = 6 + 1$

   $3 \times n + 1 = 7$

5. $d = 48$

   $d + 5 = 48 + 5$

   $(d+5) \times 3 = (48+5) \times 3$

   $(d+5) \times 3 = 159$

6. 6.2; 62.0; 620.0

7. 57.2; 572.0; 5720.0

8. 7200.0; 0.072; 0.72

9. 68

10. 71

Lesson 62 Warm Up

1. $y = 5.31$

2. $w = 0.16$

3. $n = \frac{7}{2}$

4. $m = \frac{1}{7}$

5. $w = 7.02$

6. $s = 1.08$

7. $\frac{2}{3} \times n = 17$; $n = \frac{51}{2} = 25.5$

8. p is the original price

$0.6 \times p = \$57$

$\frac{6}{10}p = \$57$

$p = \frac{10}{6} \times \$57 = \$95$

9. $\frac{49}{64}$ or 0.765625

10. $\frac{52}{3}$ or $17.\overline{3}$ or $17\frac{1}{3}$

Lesson 63 Warm Up

1. $n = 45$

2. $m = 32$

3. $38 + (^-19) = 19$

4. $84 + (^-36) = 48$

5. $k - 37 = 84$

$k = 121$

6. $72 - h = 18$

$h = 54$

7. $\frac{12}{7}$

8. $\frac{1}{18}$

9. $\frac{100}{64}$

10. $\frac{4}{1}$

Lesson 64 Warm Up

1. $^-11$

2. $^-42$

3. 6

4. 64

5. $8 - 2$

6. $m = 19$

7. $w = 4$

8. $k = 2$

9. $k = 48.6$

10. $w = 3.6$

Lesson 65 Warm Up

1. Area $= ^-(11 \times 4) + (14 \times 8) = 68$

2. $m = 12$

3. $d = 13$

4. $(\$10 - \$2) + (\$12 - \$2) = $ d

5. $p - 10 + 50 = 64$

6. $f - 3$

7. $^-9 + 2$

8. $6 + d$

9. 56

10. $^-18$

Lesson 66 Warm Up

1. $^-51$

2. 38

3. $^-81$

4. $c = ^-18$

5. $f = ^-16$

6. $^-19$

7. 100

8. 15,275

9. 15,925

10. 13.3534

Lesson 67 Warm Up

1. $1 - (^-7) = 8$

2. $3 - (^-3) = 6$

3. $7 + (^-8) = ^-1$

4. $13 + 5 = 18$

5. $^-6 + 8 = 2$

6. 32

7. 18

8. $g = 14$

9. $\dfrac{60 \text{ miles}}{1 \text{ hour}}$

   $\dfrac{60 \text{ miles}}{1 \text{ hour}} \times 4\frac{1}{2} \text{ hours} = 270 \text{ miles}$

10. $\dfrac{2.8 \text{ beats}}{1 \text{ second}}$

    $\dfrac{60 \text{ seconds}}{1 \text{ minute}}$

    $\dfrac{2.8 \text{ beats}}{1 \text{ second}} \times \dfrac{60 \text{ seconds}}{1 \text{ minute}} = \dfrac{168 \text{ beats}}{\text{minute}}$

Lesson 68 Warm Up

1. $=$

2. $=$

3. $n =$ money in Mandy's account after 12 weeks in summer

   $\$825 - \dfrac{\$25}{1 \text{ week}} \times 12 \text{ weeks} = n;\ n = \$525$

4. $^-140$ dollars

5. 12 inches

6. $m = 8$

7. $t = 14$

8. $\dfrac{20}{21}$

9. $\dfrac{43}{36}$ or $1.19\overline{4}$

10. $\dfrac{43}{9}$ or $4.\overline{7}$

Lesson 69 Warm Up

1.

2. $^-3\frac{5}{8} = ^-(3 + \frac{5}{8}) = ^-3 + \frac{^-5}{8} = \frac{^-24}{8} + \frac{^-5}{8} = \frac{^-29}{8}$

3. $7\frac{1}{9} = 7 + \frac{1}{9} = \frac{63}{9} + \frac{1}{9} = \frac{64}{9}$

4. $7\frac{99}{100} = 7 + \frac{99}{100} = \frac{700}{100} + \frac{99}{100} = \frac{799}{100}$

5. $8, ^-8, ^-8, 8$

6. $142.5, ^-142.5, 142.5, ^-142.5$

7. 6.25

8. 1.75

9. $\dfrac{19}{12}$ or $1.58\overline{3}$

10. $\dfrac{47}{8}$ or 5.875

Lesson 70 Warm Up

1. a. rational number

   b. rational number, integer, whole number

   c. rational number

   d. rational number, integer

2. $\dfrac{43}{4}$ or 10.75

3. $\dfrac{100}{32}$

4. $\dfrac{7}{8}$

5. $\dfrac{1}{1}$

6. $\dfrac{13}{1}$

7. $k = 1.3$

8. $d = 12$

9. $t = \dfrac{1}{15}$ or $0.0\overline{6}$

10. $s = 1,572,400$

Lesson 71 Warm Up

1.

n	f
⁻4	⁻3
0	0
2	$\frac{3}{2}$
4	3

2.

g	d
⁻1	6
0	0
1	⁻6
2	⁻12

3. $\frac{3}{8}$

4. $\frac{-10}{6}$

5. $k - 12 = \frac{5}{4}$

$k = \frac{53}{4} = 13.25$

6. $k \times 3 = 39$

$k = 13$

7. $\frac{5}{6}$ or $0.8\overline{3}$

8. $\frac{1}{4}$ or 0.25

9. 47

10. $\frac{1}{10}$ or 0.1

Lesson 72 Warm Up

1. $\frac{1}{2}d = 16$

2. $2 \times d = 7 + 10$

3. $6 \times d = 54$

4. $3 \times (d + 8) = 33$

5. not equivalent

6. equivalent

7. equivalent

8. $n + 7 = {}^-3$

$n + 7 + {}^-7 = {}^-3 + {}^-7$

$n = {}^-10$

9. $2 \times m = 7 + {}^-11$

$2 \times m = {}^-4$

$\frac{1}{2} \times (2 \times m) = \frac{1}{2} \times ({}^-4)$

$m = {}^-2$

10. $\frac{h}{5} = {}^-10$

$\frac{h}{5} \times 5 = {}^-10 \times 5$

$h = {}^-50$

Lesson 73 Warm Up

1. Multiply both sides by $\frac{1}{5}$, inverse property of multiplication, multiplicative property of 1, evaluate

2. $4x = 15$

$x = \frac{15}{4}$ or 3.75

3. $9 - p = 15$

${}^-9 + 9 - p = {}^-9 + 15$

${}^-p = 6$

$p = {}^-6$

4. $\frac{7}{8} \times y = \frac{1}{100}$

$\frac{7}{8} \times \frac{8}{7} \times y = \frac{1}{100} \times \frac{8}{7}$

$y = \frac{8}{700}$ or 0.01142857

5. $2 \times k + 1 \times k = 24$

$2k + 1k = 24$

$3k = 24$

$k = 8$

6. No

7. 20 feet

8. $88

9. 2,118 words

10. 2.25 ounces

Lesson 74 Warm Up

1. $n = 32$

2. $k = 10,800$

3. 25 hours

4. 19,440 beats

5. $36

6. $10k - 5j$

7. $6cde$

8. $\frac{17}{18}(s+t)$

9. $12 \times b$

10. $\frac{1}{20} \times s + \frac{3}{20} \times t + \frac{2}{20} \times u$

Lesson 75 Warm Up

1.

Tables	Chairs
2	3
4	6
10	15

2.

Minutes	Dollars
10	2
20	4
60	12

3.

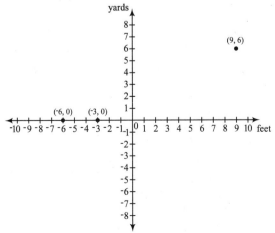

4. $x = 7$

5. $m = 18$

6. $x = {}^-1$

7. $m = 15$

8. $=$

9. $=$

10. $\neq$

Lesson 76 Warm Up

1. and 2.

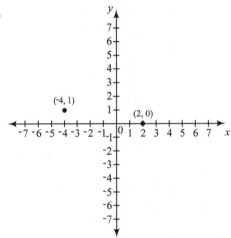

3.

m	n
$^-1$	$-\frac{3}{4}$
0	0
2	$\frac{3}{2}$
12	9

4. $d\left(\frac{5 \text{ cans}}{2 \text{ days}}\right) = c =$ cans of soda

   If $d = 2$, then $c = 5$ cans of soda

   If $d = 4$, then $c = 10$ cans of soda

5. $m\left(\frac{3 \text{ feet}}{\$1.50}\right) = f =$ feet of rope

   If $m = \$1.50$, then $f = 3$ feet of rope

   If $m = \$3.00$, then $f = 6$ feet of rope

6. $^-10$

7. $^-2$

8. 2

9. $-\dfrac{3}{14}$

10. $\dfrac{6}{7}$

Lesson 77 Warm Up

1.

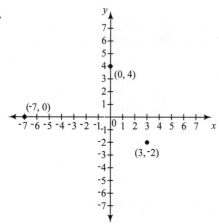

2.

m	n
$^-3$	$^-2$
0	0
3	2
6	4

3.

d	y
$^-2$	10
0	0
2	$^-5$
8	$^-20$

4. $n = 13$

5. $p = 7$

6. 12

7. $^-3$

8. $p = $ original price

$0.50p = \$47.38$

$p = \$94.76$

9. $i = $ inches

$i = \dfrac{4 \text{ inches}}{3 \text{ seconds}} \times \dfrac{60 \text{ seconds}}{1 \text{ minute}} \times \dfrac{60 \text{ minutes}}{1 \text{ hour}}$

$i = 4{,}800$ inches

10. $c = $ charge

$c = 3 \text{ hours} \times \dfrac{60 \text{ minutes}}{1 \text{ hour}} \times \dfrac{\$0.45}{\text{minute}}$

$c = \$81.00$

Lesson 78 Warm Up

1. 10

2. 11

3. 33

4. 148

5. $m = 5.5$

6. $g = 45$

7.

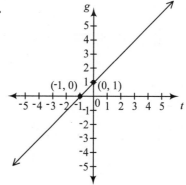

8.

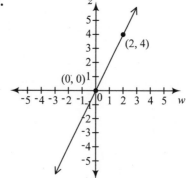

9. rate = $\dfrac{\text{1 batch of cookie}}{\text{2 pounds of flour}}$

offset = 0

10. rate = $\dfrac{\$60}{\text{1 hour}}$

offset = $25

Lesson 79 Warm Up

1. Runner 2

2. Slope = 1

3. Slope = 0

4. Slope = 3

5. Slope = $-\dfrac{2}{7}$

6. $v = 11$

7. $d = {}^-5$

8. $n = {}^-14$

9. $w = 8$

10. $x = 4$

Lesson 80 Warm Up

1.

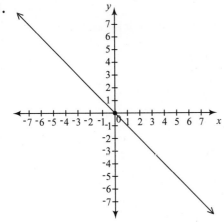

2. Slope = $\dfrac{1}{2}$

3. $k = {}^-5$

4. No

5. $c = 51$ clowns

6. $m = 5$

7. $m = 19$

8. 14

9. 2

10. 5

Lesson 81 Warm Up

1. $\dfrac{1}{12}$

2. $\dfrac{\text{30 tulips}}{\$50}$

3. (2,0)

4. $({}^-2,{}^-5)$

5. change in x is 5 , change in y is ${}^-6$

6. change in x is 8 , change in y is 2

7. 0.52 , 52%

8. 4.4 , 440%

9. 0.004 , $\dfrac{4}{1000}$

10. 4.76 , $\dfrac{476}{100}$

Lesson 82 Warm Up

1.

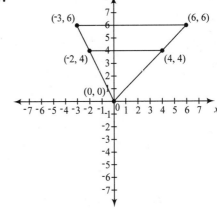

2.

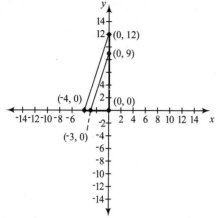

3. $(1,1)$

4. $(^-15,0)$, $(15,0)$

5. 24%

6. 2.4%

7. $\frac{35}{8}$ or 4.375

8. $\frac{72}{5}$ or 15.2

9. $\frac{75}{8}$ or 9.375

10. $\frac{213}{20}$ or 10.65

Lesson 83 Warm Up

1. $s = 1.6$

2. $m = 205$

3. 15

4. Slope $= 3$

5. Slope $= -\frac{7}{6}$

6. 24 feet high

7. 2.4

8. 60

9. 224

10. 21

Lesson 84 Warm Up

1. $y = \frac{1}{3}x$

2. $y = \frac{^-3}{4}x$

3. $y = \frac{^-5}{4}x$

4. $y = {}^-6$

5. $y = \frac{7}{6}$

6. 17 feet

7. 7 dollars

8. 2,085 words

9. Yes

10. Yes

Lesson 85 Warm Up

1. $4.80

2. 14 tables

3. $\neq$

4. $=$

5. Yes the graph goes through the origin , $y = \frac{5}{7}x$

6. Graph does not go through the origin

7. Height $= 9$

8. Height $= 14$

9. $b = $ bottles of water

$b = \frac{6 \text{ bottles}}{\$2.50} \times \$10$

$b = 24$ bottles

10. $x = $ miles train travels

$x = \frac{150 \text{ miles}}{3 \text{ hours}} \times 2 \text{ hours}$

$x = 100$ miles

Lesson 86 Warm Up

1. $\frac{6}{8}, \frac{9}{12}, \frac{12}{16}$

2. 2, 3, 5, 7, 11, 13, 17, 19, 23, 29

3. $2 \times 2 \times 5 \times 5$

4. $2 \times 2 \times 2 \times 2$

5. $3 \times 3 \times 5 \times 5$

6. No

7. Yes, $\sqrt{81} = 9$

8. No

9. $n = 144$

10. $y = 8$

Lesson 87 Warm Up

1. No

2. 2, 3, 5, 7, 11, 13, 17, 19, 23, 29

3. $2 \times 2 \times 2 \times 2 \times 2 \times 2 \times 3$

4. $2 \times 2 \times 2 \times 2 \times 3 \times 3 \times 5 \times 5$

5. No

6. Yes, $\sqrt{196} = 14$

7. $n = 9$

8. $a = 6$

9. 169

10. 225

Lesson 88 Warm Up

1. 9

2. 27

3. 100

4. 6,074

5. 13^4

6. 7^7

7. 10^3

8. 10^6

9. $e = \frac{1}{12}$

10. $g = 10$

Lesson 89 Warm Up

1. $11, 11^2, 11^3$

2. $10^{-1}, 10^{-2}, 10^{-3}$

3. $1 \times 10^1 + 6 \times 10^0 + 4 \times 10^{-1}$

4. $9 \times 10^2 + 7 \times 10^0 + 5 \times 10^{-2}$

5. $\frac{1}{2}$

6. $\frac{16}{9}$

7. $\frac{1}{225}$

8. 1

9. $\frac{18}{5}$

10. $\frac{1}{2}$

Lesson 90 Warm Up

1. 0.0001

2. 0.216

3. $\frac{10}{7}$

4. $\frac{4}{5}$

5. y^4

6. n^7

7. No

8. Yes; $\sqrt{10,000} = 100$

9. $d = 7$

10. $s = \frac{12}{10}$

Lesson 91 Warm Up

1. $\frac{4}{3}$

2. $-\frac{1}{13}$

3. 4 and 5

4. 5 and 6

5. $x = 0$

6. $x = 3$

7. $b = 1$

8. $b = 17$

9. 3

10. 4

Lesson 92 Warm Up

1. x-intercept $= 0$ and y-intercept $= 0$

2. x-intercept $= 6$ and y-intercept $= {}^-3$

3. $y = \frac{5}{6}$

4. $y = {}^-5$

5. $x = \frac{1}{3}$

6. $x = {}^-2$

7. and 8.

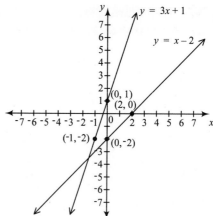

9. $y = \frac{1}{2}x$ cups

10. $y = \frac{9}{7}x$ times

Lesson 93 Warm Up

1. and 2.

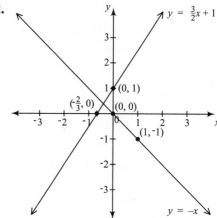

3. $x = 2, \quad y = 6$

4. $x = \frac{-8}{5}, \quad y = \frac{2}{5}$

5. $\frac{1}{2}n = 17 - 3$

6. $6n = 38 + 10$

7. $\frac{6}{5} > \frac{5}{6}$

8. $\frac{9}{9} < 9$

9. $\frac{6}{2} = \frac{81}{27}$

10. $\frac{11}{30} < \frac{12}{31}$

Lesson 94 Warm Up

1. 45 players

2. 2.5 minutes

3. 15 dollars

4. 4.5 miles

5. $y = \frac{1}{2}x - 2$

6. $y \geq x + 6$

7. $y = 7x - 10$

8. x is how much 8 boxes weigh;
   $x = 8 \times 7$ pounds
   $x = 56$ pounds

9. x is the original price of the shirt;
   $x - x \times \frac{20}{100} = \$19.20$
   $x = \$24$

10. x is the cost of 4 tulips;
    $x = 4 \times \frac{\$18}{12}$
    $x = \$6$

Lesson 95 Warm Up

1. 52

2. 25

3. 1,136

4. 999

5. $\frac{81}{100}$

6. $\frac{34}{25}$

7. $a = 0.45$

8. $r = 9$

9. 9

10. 72,000

Lesson 96 Warm Up

1. 76,350.0; 763,500.0; 7,635,000.0

2. 763.5; 76.35; 7.635

3. $2 \times 2 \times 5 \times 5$

4. $2 \times 2 \times 2 \times 5 \times 5 \times 5$

5. 0.5

6. 0.25

7. $0.1\overline{6}$

8. 0.125

9. $d = 2.7$

10. $8\frac{3}{7}$

Lesson 97 Warm Up

1. $y = 50$

2. $k = {}^-6$

3. $d = 5$

4. $z = \frac{2}{17}$

5. No

6. Yes

7. $12t + 12 = 55$ (answers may vary)

8. a. 0.2
   b. $0.1\overline{6}$
   c. $0.\overline{1}$

9. $\frac{3}{8}$ is a terminating decimal because the denominator, 8, only has prime factors of 2.

10. $\frac{8}{3}$ is a repeating decimal because the fraction is simplified and the denominator has a prime factor of 3.

ANSWERS

Lesson 98 Warm Up

1. 2

2. $2\frac{2}{3}$

3. $x = 2$

4. $x = 15$

5. $x = 27$

6. $x = 30$

7. $2y + 5 = 17$

8. $\frac{1}{2}y = 60 - 10$

9. x is the desired number;
$x \times \frac{16}{100} = 44$; $x = 275$

10. x is the regular price
$x \times \frac{40}{100} = \$36.42$; $x = \$91.05$

Lesson 99 Warm Up

1. $3x + 7x$

2. $a(b+c)$

3. No

4. Yes

5. Yes

6. No

7. $x = {}^-2$

8. $b = \frac{1}{5}$

9. x is the number of pages in the midterm
$x + 10 = 18$; $x = 8$

10. x is the price of a small pizza
$2x - 3 = 15$; $x = 9$

Lesson 100 Warm Up

1. 3^4

2. 3^{-1}

3. 3^{-2}

4. 3^0

5. 0.01

6. 0.25

7. $u = \frac{3}{11}$

8. $q = 7$

9. $4x = 13$;
$x = \frac{13}{4}$ or $3\frac{1}{4}$ or 3.25

10. $x = 43$

Lesson 101 Warm Up

1. $\frac{1}{7}$

2. 30

3. $\frac{5}{6}$

4. $\frac{100}{64} = \frac{25}{16}$

5. $\frac{1}{xy}$

6. $\frac{vt}{u}$ or $\frac{tv}{u}$

7. 1.5

8. 0.7

9. $x = \frac{2}{3}$

10. $x = 7$